Architectural Construction Drawings with AutoCAD® R14

ARCHITECTURAL CONSTRUCTION DRAWINGS WITH AUTOCAD® R14

James Snyder

John Wiley & Sons, Inc.
New York • Chichester • Weinheim • Brisbane • Singapore • Toronto

This text is printed on acid-free paper. ∞

Copyright © 1998 by John Wiley & Sons, Inc.

All rights reserved. Published simultaneously in Canada.

No part of this publication may be reproduced, stored in a retrieval system or transmitted in any form or by any means, electronic, mechanical, photocopying, recording, scanning or otherwise, except as permitted under Sections 107 or 108 of the 1976 United States Copyright Act, without either the prior written permission of the Publisher, or authorization through payment of the appropriate per-copy fee to the Copyright Clearance Center, 222 Rosewood Dr., Danvers, MA 01923. (508) 750-8400, fax (508) 750-4744. Requests to the Publisher for permission should be addressed to the Permissions Department, John Wiley & Sons, Inc., 605 Third Avenue, New York, NY 10158-0012, (212) 850-6008, E-Mail: PERMREQ@WILEY.COM.

Library of Congress cataloging in Publication Data:

Snyder, James E.
    Architectural construction drawings with AutoCAD R14
/ James Snyder.
        p.   cm.
    Includes index.
    ISBN 0-471-18418-7 (pbk. : alk.)
    1. Structural drawing.   2. Architectural drawing.   3. AutoCAD
(Computer file).   4. AutoCAD LT for Windows.   I. Title.
T355.S59   1998
720'.28'402855369—dc21   97-46730
                                                            CIP

Printed in the United States of America

10 9 8 7 6 5 4 3 2

CONTENTS

CHAPTER SEVENTEEN   Details, Schedules, Notes, Large-Scale   293
                                             Plans, and Elevations

PREFACE

CAD Encounters Of The Worst Kind

What was your first encounter with AutoCAD in a real production environment? For many of us, we were shown a computer workstation, and were told it was our's to learn, and, oh by the way, we need drawings any day now. No training, no instruction manuals, and little peer support. Of course we got ourselves into this mess because we volunteered, or answered "yes" when asked if we "knew" AutoCAD. I 'm a worst case, and this book is my way of sharing my knowledge gained from making every concievable CAD mistake, so that you will not need to do the same.

In 1984 I worked for a 160-person design firm called Swimmer Cole Martinez Curtis. One day I was called into the office of the CEO, Milt Swimmer, and told almost without preamble, "Jim, I want you to find us a CAD system." I was a vice president and design director at the time, running two, often three, large department store projects at once for the Retail Design Division, and I really needed something to fill up my remaining free time. Realizing I might never see my family again, and being the fool that I am, I said, "OK, how soon do you need it?"

To be fair, I had seen this coming. I even hoped for the day when it would happen, and feared that the conservative management of the firm would wait until it was too late to play catch-up. In 1983 I had joined the Association for Computing Machinery after attending the Siggraph convention, which was held at Cal Tech in Pasadena (that's how small Siggraph was then). It was there that I met Kai Krause, who was demonstrating a 3D drawing system running on an S100 minicomputer system driving a stroke vector CRT display, a cutting edge system if ever there was one. The IBM PC, just introduced two years before, was considered too slow for "serious" graphics work.

I was the first person in the company to actually own a personal computer, and had dreamed of being able to draw on computers since the 1970s, long before I was able to finally buy one with graphic capabilities. By the time 1984 rolled around, I had learned to program and had written an interactive interior design budget estimation system for both the Commercial (Office Design) and Retail divisions of the company. The introduction of Visicalc terminated that effort, but I had gained some valuable knowledge. However, I still couldn't draw easily on the computer, except in pixel (screen dot) scale, which didn't very well translate into drafted scale.

My desire to draw and model on computers pushed me to attend the 1983 San Francisco Computer Faire (yes, that's how the technohippies spelled it), where I first encountered the IBM PC XT with its incredible 500 mb hard disk, and AutoCAD, running on the same hardware and offering 80 percent of the features of the "workstation" class CAD systems at 10 percent of their price ($2,000). I realized then that some form of CAD was financially within the reach of even small architectural firms.

It was the advent of AutoCAD that introduced a real and affordable drafting tool to the A/E community. At SCMC we started with one PC XT, a CalComp pen plotter, and AutoCAD Version 1.8. We did nearly six months of experimentation before we were ready to add more workstations and run projects on the "system." By 1986 we had over 10 million square feet of office building space on five workstations, and were approaching the same figure for department stores. If I live to be two hundred years old, I will never recover the sleep lost during those two years.

I'm a designer by avocation and training, not a computer scientist. I can no more open an assembler and hack a Basic Input-Output System (BIOS) than I could engineer a space shuttle. Instead, I use the Computer Aided Drafting tools available to get my work done. What we have been given through the efforts of all the founders of Autodesk, especially John Walker, is a really nifty set of tools for producing drawings of buildings on computers, and I have been making my living using computers as drawing tools for over thirteen years.

In those thirteen years, CAD capability has spread from the large firms which could afford minicomputer systems (Integraph or CADAM software running DEC VAXes or IBM hardware, or ComputerVision turnkey systems), to individual architects operating out of their homes. This is a true **technological revolution**. AutoCAD LT 98 has 90 percent of the functionality of Release 14 for a fraction of the price, and is light years beyond what was state of the art in Mainframe 2D CAD in 1985.

Yet we still have a long way to go. This book will show you how to get drawings produced as best we can today. Tomorrow will be better. I can say that with certainty, because the hardware and software tools we use continue to improve, often to the point of testing our ability to adapt to the changes. The continual struggle to learn rapidly improving tools adds tremendous stress to all professions that use CAD, and this has created strains throughout the architectural profession. I have written this book to help reduce the stresses in some small way, especially for the individual drafter.

I owe a great deal to all the people I worked with and whom I trained in AutoCAD at Cole Martinez Curtis and Associates. Without the effort put out by all the people I trained, coached, and sometimes counseled, without their belief in what I was trying to communicate, CAD implementation would never have succeeded at my former employer. I could single out the most noteworthy, many of whom are no longer with the firm, but all of you know who you are. Without the faith and support of those people, I would not be where I am today. We all stand on the shoulders of others; sometimes we are lucky enough to get paid for it too. This book is dedicated to everyone who helped increase my knowledge about AutoCAD, whether they were my superiors or were supervised by me.

DRAWING
MANAGEMENT

PART ONE OVERVIEW

PRODUCTIVITY

Contents

 This book is about productivity, specifically about increasing your own productivity in designing buildings and drafting their construction drawings with AutoCAD. Productivity is necessary for profit to be produced. Without profit, no architectural practice can survive for long. In the following pages you will learn how to use AutoCAD's new and old tools to increase your drawing productivity, and thereby increase your firm's profit, thus enhancing your value to the firm you work for.

3

Most architects are not accustomed to thinking very much about productivity. Those of you in the minority who do may congratulate yourselves, since you get to keep more of the revenue from your work than the rest of the profession. I make this observation from my own experience of twenty years of work in the field. I suspect that the aversion to thinking about productivity comes from the craft orientation of the architecture profession, where making drawings has often been regarded as analogous to making the building itself. It is time for that attitude to change; CAD has removed the craft from the drawing process forever, in the sense that drawings are a direct reflection of the manual craft of their producers.

Barriers to Productivity

Lack of Proper Training

It is not true that doing drawings on CAD systems does not require great skill, even though the manual craft has been removed. In fact, not only will drawing unskillfully on a CAD system produce drawings riddled with errors and inaccuracies, but lack of CAD skill can eat up so much time that no fee will be large enough to rescue a "CAD-damaged" project. The biggest culprit in the creation of CAD-damaged drawings is the poor quality of training, or the utter lack of it, in our industry.

Poor training produces AutoCAD users who are not trained to do their jobs. Instead, they are trained in the execution of the AutoCAD commands, on the assumption that they can apply commands to their work, and simply get on with it. CAD software, however, drastically changes the nature of drafting work, and requires that a person learn an entirely new mental and physical paradigm. If you feel that your training has been less than adequate, whether you are new to AutoCAD, or a veteran DOS user struggling to adapt to the new world of Windows and Release 14, take heart—you are not alone. This book is intended to help you get your work done by serving as both a tutorial and a reference to the simplest and fastest ways to draw specific parts of buildings, and to help you organize the structure of CAD drawings in ways that will make your life easier and your work more efficient.

The Productivity Price of Drawing Errors

Drawing errors on CAD drawings take many forms. Some occur in the final output of plots, where vital information is turned off, or where objects that should not appear in the plot remain visible. More common are simply badly drafted walls and windows that overlap, don't join, or have missing components.

An error in a CAD drawing will result in an error in the dimensions. If the error is in an overall dimension, all the dependent dimensions will still add up, but all will be incorrect because they are based on an incorrectly drawn object. If the error is not gross enough to be detected on a check plot, it may not be noticed until it shows up as an error in construction.

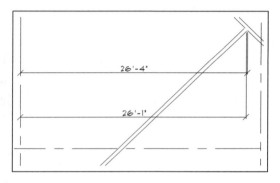

The illustration on the right shows how this works. The top dimension is the correct one, and the lower one is off by more than 3 inches. The fault in the drawing is that there is not a true intersection where the walls join, or where the grid lines cross. The dimension was done with the **Intersection** Object Snap and **Endpoint** Object Snap set as the default (running Osnap, in ACADspeak), and the drafter was working in too large a view to notice that the cursor pick box was not covering both grid lines, and that it picked the Endpoint of the wall line. The correct dimension was drawn with the default (running) Object Snap set to find Apparent Intersections.

This is the second most significant barrier to CAD productivity: the ease with which errors can be drawn, and the difficulty in detecting them. As we have seen from the above example, errors sometimes result from the tools used (in this case the Osnap settings). As I stated earlier, an error-filled drawing can take longer to correct than there is time in the schedule or profit in the fee.

The new Object Snap implementation in Release 14 goes a way toward helping us avoid errors such as the one just described. Now called AutoSnap, it gives the user the option of allowing AutoCAD to display a marker at the nearest point to the cursor, and will display a Snap Tip with the name of the point (Intersection, Endpoint, etc.). If the Osnap box covers more than one point, the tab key can be pressed to cycle through the points until the desired one is shown. When the desired Snap Tip name is visible, only that point can be selected by pressing the pick button on the mouse. The following illustration shows AutoSnap in action in the previous example drawing.

Release 14 introduces more new tools to help increase productivity than any version of the program since Release 11. All these new tools, and the older ones that often get overlooked, are summarized later in this chapter. Before we get caught up in the details of the drawing process and tools used, let's examine the larger issues of CAD drafting efficiency itself. Then we will see how the right tools, properly applied, can yield the best results.

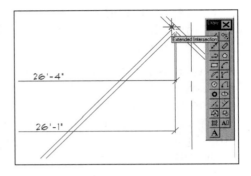

The Truth about Efficiency in CAD Drafting

Since architects began using computers to create drawings, most have done so with the assumption, or at least the hope, that creating drawings with the aid of computers would be more efficient than doing so on paper. CompuServe forums have been clogged with claims and counterclaims of the increased efficiency of CAD for years. The greater efficiency of CAD has become a Holy Grail, and its attainment has become more a matter of fervent belief than a proven fact. An assumption of the efficiency of CAD ignores a host of important variables:

- Manual drafting efficiency varies widely depending on the skill of the drafter and the methods and tools employed.

- Manual drafting efficiency is dependent on the precision needed in a given drawing, and the graphic standards adhered to. Is hatching required over large areas? Are double lines used to indicate materials instead of single lines for surfaces? Are large sections of notes and schedules hand drawn or typed? The variations are endless.

- Manual drafting efficiency can be greatly increased by using modern xerography reproduction equipment to help assemble large drawings from smaller ones, or to alter large areas of plans and elevations without erasing the area to be modified.

- CAD drafting efficiency is dependent on the skill of the drafter, but also on the speed of the computer, the resolution and performance of the graphics system, and the ease of use of the CAD software.

- CAD drafting errors are more difficult to correct, drawing progress is more difficult to monitor, and hours or days of drawing time can be lost due to computer failure or operator error.

- CAD drafting is only as efficient as its slowest piece of equipment: the plotter used for putting the drawings on paper.

- The other variables of graphic standards used also impact the efficiency of CAD drafting: Large areas of hatches slow down the graphics system of any computer, and when many lines are used to indicate something that can be drawn more sparingly, efficiency is lost.

Can a skilled CAD operator using sophisticated equipment outproduce a skilled manual drafter? **The truth is there is little or no difference in efficiency** between manual and CAD drafting.

To prove this, I reproduced an experiment that I ran many years ago when AutoCAD Release 10 first became available. I had designed an arbitrary building, shown in the following illustrations, that gave no special advantages to either CAD or manual methods. Even the glass block tower on the upper left side of the building had its diameter conveniently chosen to equal my circle template.

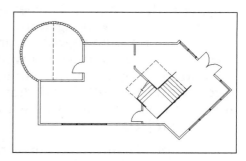

As my experimental subject I chose a drafter who is moderately fast at manual drafting and moderately fast at CAD drafting, with the edge in outright experience going to manual drafting, balanced by considerable skill on the CAD system. In other words, me.

I began by drawing the first floor by hand, ignoring sheet setup, positioning, layout, and any other extraneous issue. The goal was to measure only the time it took to put lines and arcs on paper. Next I drew the first floor on my 100 MHz Pentium computer, running AutoCAD Release 14, on Windows 95. Separate layers were used for walls, doors, stairs, and windows/glass block. I repeated the process on the drawing of the second floor, tracing it manually, and reusing the first floor on the CAD system.

The first floor took five minutes less to draw manually than on the CAD system (forty minutes by hand versus forty-five minutes on computer).

The second floor took five minutes less to draw on the CAD system than by hand (thirty minutes by hand, twenty-five minutes on CAD).

The experiment was timed to the nearest minute. Seconds really don't matter in the real world, and I didn't count them. These results are actually better for the CAD system than those produced by Release 10 several years ago: That CAD drawing took 25 percent longer than the manual one, and I used several AutoLISP programs as drawing aids to increase Release 10's efficiency at the time.

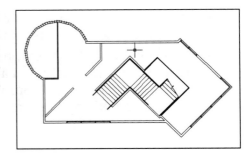

Increasing Productivity in Drawing Production

Staying with an AutoCAD Release you are familiar with under the illusion that you are working at maximum efficiency and that upgrading to a new version would degrade your efficiency during the time required to learn the new version, which would result in counterproductive output on current projects, is just that: an illusion. The new features that directly affect architectural drawings are not numerous, and are not mandated. You can still stay with the old methods for the duration of existing projects.

For example, I continually run across drafters who insist on typing <u>W</u> for Window and <u>C</u> for Crossing selection methods even though that requirement vanished with the introduction of Release 12. AutoCAD still supports this arcane behavior, even though all drafters would be more efficient if they were forced to use the (new) implied right to left for Crossing and left to right for Windows conventions introduced with Release 12.

Drafting efficiency alone doesn't result in overall productivity. There are so many factors affecting productivity in the production of design development and construction drawings that it is impossible to sift out a meaningful combination. Some factors that affect productivity can't be controlled by the architect, such as the client's requiring cumbersome layering schemes, complex drafting formats, tedious custom menus linked to block libraries, or prohibitions against reference drawings.

Other productivity killers may have to be tolerated for economic reasons: antiquated and slow networks, slow plotters, and outdated computers. The local labor market may put employees of a higher skill level out of reach both financially and physically. The cost of good training may be too high, even for in-house efforts.

But there are other strategies you can pursue that will result in dramatic gains, not just in CAD drafting, but in manual drafting as well.

Overall Productivity-Enhancing Actions for Both CAD and Manual Drawings

To start with, analyze the state of your drawings to see what can be simplified without reducing the information conveyed to the building trades. Can notes be substituted for hatches, schedules for details?

For example, look at your framing drawings. Are you drawing every joist instead of pulling down a dimension line and typing "25 x type open web joists on equal centers" and hitting the enter key? Are you drawing every stud in an exterior framing elevation? Put in a detail key referencing a typical stud-framing drawing instead, and draw only the nontypical stuff. Done. Go on to the next part of the drawing.

Do you draw every piece of steel you can imagine in a nonstructural drywall design section? Forget it. Put in a key to a typical framing drawing and leave the steel out. The framing contractor won't follow your drawing anyway, and the building department inspector can easily approve your typical drawing details.

There seems to be a tendency in the architectural profession to be obsessed with **drawing everything we know about a building**. If there is some means of graphically representing every mortar joint, every fastener, every tile and shingle, someone will try to draw it all. I don't know why this is so, but it is this lack of discipline that will suck all the hours out of a project schedule in the blink of an eye. There are parts of any building that must be drawn to be understood, and those are the items that drawing time absolutely must be concentrated on. Drawings and notes just convey information.

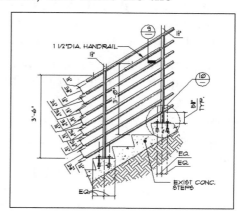

Here is a stair detail drawing that illustrates this point. It is a very pretty drawing, but it can be improved through simplification. The following elements should be dispensed with:

1. Drawing redundant objects such as the intermediate rails.
2. Highly detailed drawing of small objects such as the bolts.

3. Shading of the rails and posts. The diameter dimensions and sectioned ends of the tubes clearly indicate they are round and not square.

4. Multiple identical dimensions of identical objects.

5. Existing objects that cannot be known (soil formation, concrete depth) are hatched in detail. This makes it look like the architect is taking responsibility for existing parts of the work—a bad idea when liability is considered.

This drawing suffers from art at the expense of communication. Because of the drawing's graphic density, details are required to show parts that could otherwise be dimensioned and explained by note. The next illustration shows how this drawing can be pared down and communicate more information at the same time.

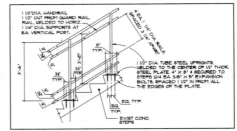

The quality of the information is the critical thing, not the quantity of graphic elements used to convey it. I am convinced that many people substitute drawing for thinking. Drawing without thinking is easy. It's understanding everything about what you are drawing that's difficult.

CAD Drafting Changes the Construction Drawing Process

Productivity can be increased by these types of changes to your drawings, but CAD drafting demands new **project process organization**, because it is exceedingly difficult to check CAD drawings while they are in progress. When drawings are done manually, it is possible to walk around the office and take a look at what's on the boards. A drawing on paper looks like the one shown above.

At right is what the same drawing looks like in the computer of a drafter who has gone home for the day.

When drawings are inside computers, it's impossible to look at them and see complete drawings at once, as we can with a paper drawing, and it may be impossible to look at them at all if someone else is accessing a particular file. It is also difficult to know where the latest version of a drawing is on a computer network. Document management software can help solve the latter problem, but only regular scheduled check plot reviews can address the former.

Get Control of the Drawing Review Process

Review of check plots is imperative because large drawings (typical of architectural plans) cannot be accurately checked on a computer monitor screen. You may think you can do it, but you will always miss something the trained eye would see on a paper plot. I have proved this assertion to my satisfaction (or chagrin) more times than I care to remember.

Check plot production and review must be scheduled so people doing the drawing can plan for it. It must be enforced, even if the job captain is out of town when the plots are scheduled to be run. Part of the check plot discipline is checking every drawing, whether anyone thinks it has been changed since the last review or not. Unfortunately, running check plots is at least as disruptive as running check prints of hand-drafted drawings, because someone forgets to turn off some layers, or doesn't set line weights properly, and has to run the plot again and again. In the chapters on plotting and drawing review, we will show how to use AutoCAD's VISRETAIN system variable, combined with well-structured reference drawings, to control plot production smoothly and automatically.

Design Development Productivity

In the design development phase, designers must learn when certain drawings need to be done in relation to others, because the geometry of an elevation can help generate a plan, and vice versa. As mentioned earlier, designers need to start thinking dimensionally from the beginning: How long is this side of the structure? How high will the codes and site conditions let me build? Planning and zoning requirements must be incorporated from the beginning to avoid redrawing. Discovery of regulatory and legal requirements for the site should be completed before the design is even sketched on the back of the proverbial envelope or napkin. The number one rule of CAD design development is: **Never draw something without exact intent.**

This is hard for many designers. I know. I'm one of them. I'll think, "Let's locate the exit stairs at the building perimeter" instead of "I'd better check the occupant load on these two floors to see if the number and size of the stairs are going to severely reduce the amount of window area." When the size and required number of exit stairs are known, and their location roughed out, go ahead and draw them in exactly as required, to actual size. It won't hurt as much as you think. It also insures that someone else won't waste time drawing them all over again and hating you for it. Critical and/or mandated parts of a building should be precisely drawn early during design development to avoid drawing them often.

This is work process change at an individual level and is illustrated in the process used in the development of the Tutorial Building drawings in this book.

How Using Computers Affects Drawing Management

The very use of computers forces changes on how drawings are stored, retrieved, and printed and determines who has access to them. Where once drawing management was a job captain responsibility, computer use pushes this work down to the individual drafter level. The single biggest change from paper to computer drawing storage is that paper drawings never have to be **SAVEd** (you just don't throw them in the trash). Only by using computers have we introduced the ability to delete hours of work in the blink of an eye. In Chapter 2 we will detail how to cope with these work process changes, specifically in drawing management.

Don't Be Lured into Believing the Drawing Automation Myth

A critical mistake in understanding CAD in the past was thinking that using a computer could somehow **automate the drawing process**. The hope was that after the design had been sketched out, the "fuzzy" wishes on form, structure, and materials could be turned into construction documents with a few keystrokes. While few architects have this illusion today, many still think that CAD will make construction drawing "easier" or "more efficient" in some undefined way. The **wish** for automation is still with us.

The myth still surfaces on CompuServe forums and in the print media from time to time. It is kept alive by third-party software producers who advertise "automated" section and elevation creation afforded by their products, and by media pundits who look at the mechanical design part of the CAD universe and think that if you can generate complex 3D models of car engines, then creating models of buildings should be an easy logical extension of the tools already available.

None of these people has ever tried to actually model a complete building in 3D and extract the necessary 2D views and sections from it, and for a good reason: such a model is too complex to actually build. Buildings contain thousands of components, many of which are extremely geometrically complicated (furniture, sprinkler systems, and indoor plantscaping, for example). I have never seen anyone model a curtain wall that could be sectioned into **construction details**, complete with glazing and framing extrusion sections specific to a manufacturer.

Existing 3D modeling programs, including those in the $10,000-plus range, can't handle this level of complexity. They certainly can't handle it on the typical computer workstation most architects can afford.

This has not stopped architects from being seduced by the myth. I recently watched a firm spend the better part of its fee creating a 3D model of a building for use as the background of a design rendering, thinking it could use it to generate the geometry for the construction drawings. The rendering was approved by the client, but then the real development of the building systems and details began. As the detail design progressed, the model quickly became obsolete, because updating it was too difficult. At a critical point in the project, the 3D model had to be abandoned, and all drawings had to be started over in two dimensions.

Ironically, the exterior design remained stable, with most of the revisions affecting the interior systems. This firm effectively drew the building twice for one fee (even with the income for additional services required by the changes to the interior design included). Of course, the original fee covered only **one** production of the building's construction drawings, not two. We will take you through a 2D approach to design development (with some quick and dirty 3D) that is intended to produce CAD files that are actually usable as the basis of construction drawings.

Once and for all, let's put this notion to rest. **CAD construction drawings are not automated!** Drawings are drawn by people, regardless of whether the tools used to produce them are manual or computerized. To draw on a CAD system, the people who will do the work need to master skills that are unique to the new drawing environment. They must also learn the system in a way that will let them work at peak efficiency and achieve increased productivity.

Making the Move from Manual to CAD Drafting

The big paradigm shift for anyone moving from manual drafting to CAD is to recognize that while manual drawings can **evolve**, CAD drawings must be **built** before they can evolve. The iterative process of design cannot follow the same timelines with CAD drawings as it does with manual drawings. Trying to use manual drawing management practices with CAD drawings will create confusion and could lead to financial disaster for your firm. There are many firms that draw drawings on computers, only to create chaos out of every project. In the process, they alienate their CAD-proficient staff, who often feel sabotaged and victimized.

Learn New Physical Skills

CAD drafting will place new demands on your body as well as your mind, but it will relieve some physical stress associated with working on a drafting board. You will make constant use of your dominant hand and arm almost exclusively. You will also have to fight visual fatigue (stress on your eyes' focusing mechanism), because you will tend to focus at a fixed distance for long periods of time. On the other hand, you will sit more comfortably and not suffer the back strain that comes from long hours of bending over the drafting board.

The major hand/eye coordination skill to learn will be the use of the mouse to move the crosshairs on the screen. This skill is not difficult to learn for someone with good small-muscle coordination—it just takes practice. If you draft, you've already got the coordination.

I always urge people to set the mouse acceleration as high as necessary to allow them to cover the entire monitor screen with one small wrist movement. This takes all the pressure off the forearm and shoulder, preventing neck and shoulder pain from continued contraction of the arm muscles. These illustrations show the amount of hand movement and the amount of screen area this maximum motion should cover:

 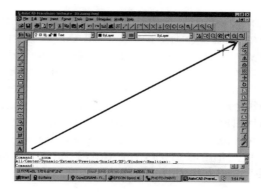

The next section applies to new AutoCAD users as well as veterans upgrading from older DOS-based versions.

Making the Change from DOS AutoCAD to Windows 95 or NT Versions of Release 14/AutoCAD LT 97

Learn the Icons

Most of the command icons have been changed from Release 12 Windows and AutoCAD LT for Windows 3.1. This is unfortunate and a major irritant for anyone expecting backward compatibility between Release 12 and 14, but the icons have not changed at all from Release 13 to 14. If you have been an exclusive DOS user, the bright side is that you are starting out even with Windows 3.1 users on the icon recognition learning cycle. Release 14 has Tool Help (pop-up names for icons appear if you pause the cursor over a button for several seconds), which was previously available only to AutoCAD LT 2.0 users. Tool Help makes the learning curve more user inclined. For those of you who still feel excessively burdened, just be glad you no longer have to deal with learning a new digitizer tablet layout without any help at all, as we did during the Release 9, 10, and 11 upgrades.

The main reason to learn the icon symbols is that pushing these virtual buttons with the mouse is far faster than entering commands any other way, on average. I know many of you are fast keyboard command veterans, but so am I. I timed both methods and discovered that even using the pull-down menus was faster than keyboard command entry on average, and pushing buttons was even faster than making menu picks. I still use keyboard entry for the ERASE command, and for REDRAW. This is largely a matter of personal preference. If you are left-handed, you may consider using L for LINE and other keyboard shortcuts available to your free hand. Just remember that keystrokes used to issue AutoCAD commands need two inputs: the key command (L for LINE, for example), and the [Enter] key for confirmation. Use the Enter button on the mouse as much as possible, because your hand is already positioned to press it, and right-handed people don't have to take their hand off the mouse and look at the [Enter] key.

Mouse Button Conventions in This Book

In all the illustrations, the left mouse button is used as the Pick button (the one you draw with), and the right mouse button is the Enter button, which is the AutoCAD default assignment for this button.

Left-handed people can reverse this easily in Windows 95 and NT. AutoCAD will happily accept the button assignments as long as the system pointing device is used in Release 14. AutoCAD LT 97 and Release 14 pick up the mouse's configuration automatically.

Keep Your Eyes on the Screen by Using the Keyboard Sparingly

Imagine drafting manually and being required to reach down and type on a keyboard or point to a symbol on a separate tablet every time you wanted to draw a line. Not only is such a requirement time wasting, it breaks your concentration on the drawing and interrupts the flow of work. This is equally true of CAD drafting. Fight old typing habits and use the icon tool buttons. You will be rewarded with speed increases and better concentration. The best thing about the tool icons is that they can be positioned anywhere you want on the screen to suit your type of work, and you can create any combination of tools you want: AutoCAD no longer controls how you interact with the computer; you are now in complete control of the user interface. You can even create your own symbols for commands if you don't like AutoCAD's.

Learn and Use the New Features Offered by Release 14

AutoSnap, Tracking, and **Direct Distance Entry**, **Running Osnap on/off control**, and the **Layer Change** button are major time savers, as is editing with Grips. We will cover the use of these new tools (OK, Grips aren't new) in the rest of the book. On the subject of Grip editing: If you are a Release 12 user who hasn't practiced using Grips instead of the MOVE and STRETCH commands, it's time you enjoyed the intuitive and direct way Grips can accomplish the things you once used these commands to do.

Take Advantage of Tool Bar/Box Customization

Place the tools you use most where you need them. The best thing about tool bars and boxes is that you can tailor the interface between you and AutoCAD to work the way you work. No longer are you forced to memorize keyboard commands or their aliases, or to customize menus to have commands conveniently placed. See Appendix A on customizing the tool bar and boxes.

If you specialize in a narrow branch of architectural practice and have created several customized menus and icon menus to insert blocks and perform other drawing operations under Release 12 or older versions of AutoCAD, you will need to recompile them in Release 14/AutoCAD LT 97. On the other hand, if you have several favorite AutoLISP programs that perform mundane drafting tasks such as layer management, take a hard look at the new features now offered by the standard software. AutoCAD and the Windows 95 interface have ways of accomplishing most of the things we used to write specialized programs to do.

AutoCAD is now almost fully equipped for architectural drafting. Of course, the makers of architectural front end programs will dispute this, but look at what we have now, as opposed to what was available in Release 11:

The Release number in which the tool or feature was introduced is listed at the far left. LT stands for AutoCAD LT 95 and 97 only. If a feature has been highly modified in Release 14, it is given the R 14 designation.

R 12	🖉	double line wall construction
R 13/LT		multi line linetypes
R 14	A	true paragraph text with a true editor, and 50 percent faster text processing than Release 13
R 13/LT	✓	spelling checking
LT/R 14	OSNAP	Object snap on/off toggle button is on the screen (forget the [F5] key).

LT/R 13		Direct Distance Entry
LT/R 14		tracking (the ability to start drawing a line or other object relative to a series of points selected prior to creating the ob1ject)
R 13/LT		associative hatching with automatic boundary detection
R 12		associative dimensioning
R 12		interactive editing with grips (instead of commands)
R 14		one-button layer switching (push the button, point to an object on the desired layer, and immediately that layer becomes the current one)
R 13/LT		on-screen layer management for freezing, thawing, and turning layers on and off without using a dialog box
LT/14		one-button transfer of a source object's properties (all or selected properties such as color, line type, layer, etc.) to any number of selected objects
R 12		icon-based command interface (on the screen instead of on the digitizer tablet)
R 13/LT		access to all True Type type fonts
R 12		dimension styles and graphic setup for dimension styles
R 12		automatic window or crossing box selection.
R 14		external reference drawings, including the ability to manage path assignment, and to clip out selected parts of reference drawings that are not needed by the drawing in which the external reference is to be inserted
R 13/LT		true ellipses and elliptical arcs
R 13/LT		NURBS-based spline curves
R 13/LT		three new selection methods (window polygon, crossing polygon, and fence)
LT/R 14		real time (interactive) pan
LT/R 14		real time (interactive) zoom

R 12	⊡	aerial views of the drawing
R 13/LT	⊡	construction lines
R 13/LT	⊡	rays (infinite lines)
R 12	⊡	copy objects to and paste in the Windows Clipboard

None of these capabilities were available in Release 11 only a few years ago. Only nine were available in Release 12. These new capabilities allow us to do some things we could only dream of in the past, or for which we had to buy third-party programs. I remember pundits in the trade journals crying "If only AutoCAD had a way to draw double line walls, it would be a great architectural drafting program!" When Release 12 came out with double lines, they demanded yet more features. The point of this is that there is no end to our desires, but in the meantime we have drawings to do.

These new features, combined with all of AutoCAD's other tools, make the program a very capable architectural drafting platform. I am ready to declare it Good Enough, as my own personal wish list is now very short. As you will see throughout the book, we can use all these new tools to develop methods for drawing that will contribute to increases in productivity never afforded under DOS.

DRAWING STRUCTURE AND ORGANIZATION

Contents

CAD drawings must be **designed** and **constructed**, just like any building project. Bad design decisions made at the beginning will lead to drawing construction disasters in the production drawing phase. While there is a parallel to this in your practice of architecture, there is little parallel to the manual drafting process. The design part of the drawing construction is the establishment of the type of relational structure and internal structure your drawings will have.

The Major Differences between CAD and Hand-Drawn Construction Drawings

Manually drafted drawings are approximations of the building to be constructed, and the drawing is meant to convey design intent only. Dimensions on manually drafted drawings are not derived from the drawing itself, but represent mathematically correct specification of the structure. A math error in the dimensions can be detected by adding and comparing detail dimension to overall numbers.

The CAD drawing, in contrast, is a full-size representation of the building. This is a revolutionary concept in the practice of architecture. For the first time since ancient civilizations began to stack stones up, we can now draw a building with more precision than that with which it can be constructed. Dimensions in a CAD drawing are annotations of the drawing's objects (lines, circles, polygons, etc.) that represent building components. If the drawn objects are not correct, the dimensions will only reflect those errors.

CAD's Impact on the Relationship between Small-Scale and Large-Scale Drawings

CAD has drastically changed the relationship between small- and large-scale drawings. Hand-drawn large-scale drawings have always been done to insure accuracy in construction of critical building components. They were intended to convey detail dimensional and assembly information not visible at smaller scales. This meant that some dimensions could not be set for things like the size of stairwells until the large-scale sections and plans were developed. It was critical to make sure that the dimensions shown on the small-scale plans were updated to reflect drawings developed at a later date.

Since CAD drawings are produced at full size from the start, all the detail dimensional information required for construction is always available in a single drawing. This means that as the design is developed in "small-scale" plans and elevations, more time must be spent than in manual drawing to pin down dimensions, and to draw exactly, rather than approximately. This requires more design decision making at a detail level during a phase of the process when designers are unaccustomed to making precise decisions.

This is a drastic change for most architects, because they are forced to consider the dimensions of everything in the building before drawing it. Such a constraint requires much more up-front work on paper before the CAD system is even turned on. The alternative, to develop the design in the CAD system and continually redraw it as the inevitable changes occur, can leave the CAD project in a state of chaos deep in the project schedule. I have developed methods for design development in the CAD system that attempt to minimize this risk.

Large-scale (plotted) drawings are still needed in a CAD project to explain assembly of small components, fit and finish requirements, and level of craftsmanship. If the CAD system can be used to integrate manufacturers' details with repeated standard drawings, time can be saved on the project. See Chapter 17 on detail drawings for shortcuts and tips on how to do this and how to use CAD to accelerate the hand drawing of project-specific details. The fact remains that trying to draw all of a building's details on CAD is not productive. At the very best, all you can hope to do is equal the time required to do the same task by hand (as we showed in the Chapter 1 section "The Truth about Efficiency in CAD Drafting").

Dealing with an Employer's Existing CAD Standards

If you work for (or even own) a firm that already has established CAD drafting/graphic standards, you may still find the following sections interesting. Feel free to use our scheme to evaluate your own systems, but don't be tempted to fix what isn't broken. On the other hand, if your firm is not making use of external (xref) drawings, and is working with more than thirty layers per drawing, our system can simplify your life a bit.

Don't Take My Word for It

If you are unsure about the usefulness of some structure in this book, give it a test against a known benchmark. If you are not making use of external reference drawings, for instance, time how long it takes to make a revision affecting more than one dependent (plotted) drawing. Break the drawings up into xref files, as we describe in the next section, and make the same revision, tracking the time used. If the time required under the new system is about the same as under your existing methods, you can be fairly sure that your productivity will improve with the new system as you gain experience with it.

Setting CAD Drawing Structure Standards

Internal drawing standards for layering (including plotted line weight mapped to specific colors), drawing format, sheet layout, and common text fonts should be decided on and put in writing, even in a single-person practice. Every practice should also establish drawing management standards specifying what directories (folders) text fonts will be located in, project folder name conventions, and file-naming conventions. Folders for common component libraries should also be specified. Taking these steps (if you have not done so) will double or triple your CAD productivity. There is nothing more time wasting than chaos in layering or the inability of AutoCAD to locate fonts every time it opens a drawing.

Once you have standards, make sure all your consultants work with them. Make doing so a condition of your contract with them. Meet to go over your standards and to get all consultants' agreement ahead of time on how they will integrate their drawings with yours. Don't be a "Nazi architect" by rigidly imposing standards. Remember, the goal of being organized within a team is to make doing the work easier for everyone involved. Bend your standards to enhance the team's organization.

I know a large architectural firm that demands that all consultants use the firm's directory (folder) names for CAD projects. Unfortunately, its directory names use multiple periods as allowed by Windows 95 and NT—for example, "Project x.3 floor 2". This firm doesn't care that consulting firms might still be using operating systems that don't recognize this name structure, and will not allow directory names with spaces and more than three characters after a period. As a result, its consultants aren't able to open its xref drawings, or find the directory containing special text fonts.

We will cover this and other issues relating to working with consultants in considerable depth in future chapters. It is important to recognize that the process an architectural project goes through is by nature chaotic. Building patterns for the work process helps to lessen the chaos and makes deviation from a productive process easier to recognize and control.

External drawing standards define how different drawings relate to one another to create different plan and elevation views of the building. The external drawing dependency structure determines how efficiently your CAD production engine will run. If you make it too complex, broken links in the dependency chain can cause major production problems for you and your consultants. I have developed a basic drawing dependency structure that is designed to avoid complex layering schemes and complex overlays, and it has the additional advantage of keeping drawing file sizes as small as possible.

Setting Up a Drawing Dependency Structure

The diagram following illustrates how we use the CAD system to generate plan drawings by using other drawings as underlays and overlays, similar to the old pin bar system in manual drafting. Drawings are overlaid on other drawings in AutoCAD by issuing the XREF command, followed by the ATTACH command, or by simply pressing the 🔲XREF button on the XREF tool box. The drawing file being used as the attached XREF can be accessed for this purpose by many people at the same time, even while someone is working on it (if it has not been locked). As changes are made to the attached file (assuming everyone is connected over a network), all of the file's users will be notified by AutoCAD that the file has been changed, and they will be asked if they want to reload the XREF. The drawing structure shown is for each individual floor of a building.

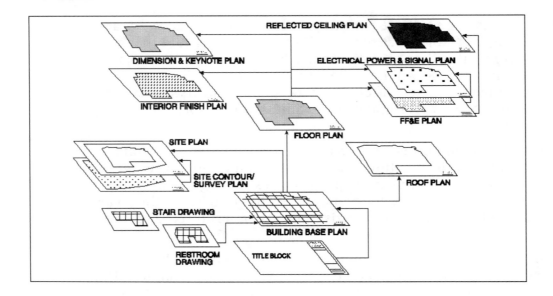

Title Block Drawing

At the "root" of the tree in our illustration is the title block drawing. It is automatically attached to all other drawings, because it is specifically attached to the next drawing in the tree, the building base plan. In practice, it is best to attach the title block drawing at the end of the construction drawing process, because title blocks contain a great deal of text and other graphics that slow drawing regeneration.

The Building Base Plan

This drawing is referenced by all other plan drawings. It is also the file that is **always** distributed to consultants. It contains the building's exterior walls, shaftways, columns and structural grid, north arrow, exterior doors, and other fixed elements such as stairs and rest rooms. As illustrated here, the rest rooms and stairs may be external reference drawings, or separate drawings that are inserted as blocks. Base plan files always contain this information and only this information.

Rest Room and Stair Drawings

On multifloor buildings, these components often repeat, and they also must be turned into large-scale plans for detailed dimensioning and finish/equipment annotation. Keeping them separate allows them to be manipulated more easily and scaled independently of the other plan drawings.

Site Contour/Survey Plan

On large sites, or sites with complex topology, it is a good idea to make a separate file for the site information. This allows you to turn off layers of objects that are being removed or altered, without having to incorporate all the complex data into one file. As the site survey drawing is updated, the impact of new conditions is easily seen on the site plan to which it is attached.

Site Plan

This file contains the property boundary and its coordinates, dimensions, and planting and hardscape improvements, and, of course, the building outline. The building outline is traced over the base plan (first floor) exterior wall, then all the base plan layers are frozen before the drawing is plotted.

Roof Plan

The roof plan uses the base plan of the top floor for reference and is traced over the building wall outline. Normally the layers of the base plan are frozen once the outline has been generated and roof drains have been located.

Floor Plan

This plan handles all interior partitions, doors, interior glazing, annotation, and furniture, fixtures, and equipment, if included. In buildings with complex FF&E, such as a hotel, for instance, that information would be developed in another file that would attach this drawing to itself. The floor plan attaches the building base plan file.

FF&E Plan

This file is created in cases where there is a great deal of graphic and text information required, such as in an office floor with many systems of furniture workstations. It attaches the floor plan. Electrified cabinet work or other equipment or workstation panels at power connection points are on a unique layer that can be made visible on the electrical power and signal plan. In the early stages of the drawing process, it may be beneficial to develop FF&E layouts on the floor plan, but split it out at a later date into its own file.

Reflected Ceiling Plan

The reflected ceiling file contains the ceiling system grid, door headers, ceiling soffits, drapery pockets, light fixtures, exit signs, HVAC supply and return diffusers, etc. It also contains all note keys for ceiling types and ratings, light fixture type, and dimensions for drywall ceiling/soffit construction or light fixture location. It attaches the base plan to itself.

Electrical Power and Signal Plan

This file contains all floor and wall power and communications outlet locations, switch locations, sensors, exit signs, strobes, etc., and dimensions. It attaches the floor plan and the FF&E plan.

Interior Finish Plans

These drawings have all paint, flooring materials, wall coverings, and other interior finish location keys. There may be more than one finish plan required; for example, wall and floor finishes often require separate files. All finish plans reference the floor plans.

Dimension and Keynote Plan

This file contains all interior partition dimensions, wall type keynotes, door type symbols, interior millwork detail keys, and other detail symbols. It attaches the floor plan. This plan does not need to be a separate file on buildings as basic as our Tutorial example. It is usually only required on very large complex structures.

Reference Drawing Insertion Points

Reference drawings should (almost) always be inserted at the 0,0 coordinate. Setting this as your office standard will avoid all kinds of problems, and it insures that anyone can call a drawing as an xref without knowing where its insert point is, because AutoCAD uses 0,0 as the default coordinate. The other standing rule that is needed to complement the first one is that no one is allowed to move the building base or floor plan relative to the 0,0 point once reference drawings have been started. Moving these plans can happen if someone needs space for something like a block of text or a legend. If the building base plan moves, it won't align with any of the underlays anymore, and you can end up with nine or ten misaligned drawings in an instant. Never make this kind of alteration to base or floor plans without the permission of the whole project team and any consultants involved.

Diagram your xref structure and distribute it to the project team members. As you begin working with consultants, give them a copy.

A Strategy for Naming Files

The storage and retrieval of data on a CAD system is its heart and soul. **Without a good file structure, a busy office has chaos.** We should strive to construct the most efficient, intuitive file system that supports the way drawings are drawn by architects. The support for long file names (up to 256 characters) in Windows 95 and NT is a true blessing, because it allows us to indicate important information such as file contents and revision status in words instead of cryptic code. Unfortunately, many of your consultants may not be using these advanced operating systems, so you will also need a way to incorporate the old DOS 8-character file name in drawings shared with them. A little later I'll offer some suggestions on how that can be accomplished. First let's examine some issues affecting file name systems, and what we want our system to accomplish.

AutoCAD's Limits on Long File Names

Multiple Periods Are Not Allowed

When AutoCAD encounters a period, it expects to see "DWG" on the right side of it. Therefore, a legal Windows 95 file name such as My house.drawing 5.1.dwg will be read as "My house.dra" by AutoCAD and you will receive an error message stating that the file is not a valid drawing. Always use a dash (hyphen) instead of a period: My-house-drawing5-1.dwg.

Spaces Create Problems

When you try to attach a drawing file with spaces in its name, AutoCAD will not allow the attachment, because a reference drawing is inserted as a block definition into the drawing calling it, and blocks can't have spaces in their names. There is a trick to get around this: Say you are trying to attach My house-drawing 5-1 to My_Site_Plan. In the Xref dialog box, in the "Xref Found At" window, type "houseplan=My-house-drawing 5-1.dwg" and press [Enter]. You must type the double quotes. To avoid this problem, use underscores or dashes instead of spaces: My_house-drawing_5-1.dwg.

As a general rule, also avoid any characters that are not allowed in DOS file names, so you won't need to rename files when sending them to consultants who use DOS or Windows 3.x.

A Prototype File Name Structure

The Four-Character Project Identifier: Our office uses the last two digits of the year and a two-digit project number as our project code—for example, 9804, which is project 4, in 1998. I consult with a firm that uses a three-letter and two-number project identifier (such as MGM-12). That kind of code would need to be shortened to something like MG12. Whatever your system, you can probably extract four characters from it that will distinguish the project sufficiently.

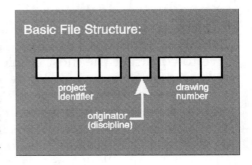

Basic File Structure:

project identifier — originator (discipline) — drawing number

The One-Character Originator/Discipline Code: This code follows the system used in the 1990 AIA Layer Guidelines, with some disciplines added.

A	Architect
C	Civil
E	Electrical
F	Fire Protection
G	Geographic Information Systems (GIS)
I	Interior Design
L	Landscape Design
M	Mechanical
O	Other
P	Plumbing
Q	Equipment
R	Reflected (ceiling or other reflected plans)
S	Structural
T	Telecommunications/Data

You're free to use your own designations, or substitute letters, of course. Another approach would be to keep different disciplines' files in separate folders. We recommend the above method, however, because it follows the convention of using the discipline code as the first character in the drawing sheet number.

The Three-Character Drawing Number Code: Many architects use drawing sheet numbers such as A-1.02, A-1.03, or A-101, A-102. A simple modification of this system allows us to create easily read file names showing us the project and the drawing number in the first eight characters: 9804A201 means that this is the first drawing of the second-floor architectural set. 9804E201 would indicate the first electrical drawing for the second floor.

In Windows 95 and NT we can expand this basic structure to provide the drawing title and other data:

9804A201_Building_Shell_&_Core_Plan-rev_7-9-98_by_JS.dwg

When we send the drawing to our electrical engineer who is still working with DOS and Release 12, we pick the Save As option on the File menu at the top screen menu bar, and in the Save Drawing As dialog box, delete everything to the right of the first eight characters. Drop down the file type list at the bottom of the box, pick Release 12, and pick OK. That will create a Release 12 file named 9804A201.dwg. This makes Release 12 versions of your files easy to spot in either My Computer or Explorer views of project folders. The use of long file names complicates the use of reference drawings. See the section of this chapter titled "Who Has the Latest Drawing?" for some suggestions on incorporating the maximum amount of information without messing up xref structures.

An Alternate File Name Structure

Here's another file name structure that has a higher level of complexity but allows more detail:

The Two-Character Client Identifier: This can use letters or numbers. If you have two clients with the same initials, you will need to be creative in picking a mnemonic abbreviation.

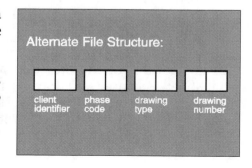

Alternate File Structure:

client identifier | phase code | drawing type | drawing number

The Two-Character Phase Code:

PO	Proposal
PR	Programming
SD	Schematic Design
DD	Design Development
CD	Construction Drawings
AD	Addendum
RE	Revision/Add Services

The Two-Character Drawing Type Code: The first letter is the discipline code from the previous scheme. The second letter is the drawing type code:

C	Cover
P	Plan
E	Elevation
D	Details
S	Building Sections
N	Notes and Schedules

The discipline code is the same as the previous scheme, but we lose some precision in the drawing number. Here's our previous **9804A201.dwg** file in this system: **IADDAP02.dwg**. The IA stands for Image Advertising, giving us the Design Development Architectural Building Base Plan for the second floor. IADDAE05.dwg would designate the Image Advertising Design Development Architectural Elevation drawing, number 5.

Principles of Drawing File Access Control and Version Tracking

Checking Files In and Out

In manual drafting, the project manager, project architect, or job captain has control of the drawings and generally keeps track of who has what sheet. Because CAD drawing files are invisible, residing on workstation hard disks or on the network server, knowing who is accessing a given file is extremely difficult. You can go over to someone's desk or to a drawer and check on the condition of a paper drawing. It is not easy to check on the integrity and contents of computer files just by looking at their listing in a folder.

Saving CAD Drawings from Ourselves

Computer files are extremely vulnerable to loss or destruction, which can happen when users are careless or hardware and software suffer failure. By far the easiest way to destroy a file (making it unrecoverable by unerase programs) is to overwrite it with a different file of the same name. It's as simple as opening a new blank drawing and saving it to the same name as an existing one. Of course you will be asked if you want to replace the existing file, and given a chance to cancel the overwrite, but it's still easy to do. Using Explorer to drag and drop groups of files from one folder or disk to another is a great way to accomplish this because you can say "yes to all" overwrites without realizing that you are replacing newer versions of files with older ones.

There is a serious bug in Release 13c4a that can overwrite one file with another. If you are in an existing named drawing and select File Open, pick a new file in the File Open dialog box, pick OK, and then immediately cancel (by pressing [Esc] or [Ctrl]+[C]), AutoCAD keeps the current drawing in memory, as it is supposed to do. This same current drawing name displays correctly in the title bar at the top of the screen, and everything looks OK. When the Open Drawing command is used, however, AutoCAD invisibly switches the DWGNAME system variable to the name you picked before canceling the File Open operation. Now any ▤ Qsave you make will overwrite the file you selected and then canceled, with the contents of the drawing on the screen. Worse yet, your current drawing file will not be updated, and you could end up with duplicate file names. Release 14 has several traps for this type of error and can warn you of impending unintentional file overwrites. Nothing in Release 14 prevents someone from using the COPY (drag and drop) command in Windows 95 Explorer to overwrite any file in any directory.

Access control of drawing files won't make careless overwrites less likely by the people who have access privileges, but it will restrict the number of people with these privileges. Access control of files will also make the people granted access aware that their privileges are explicit and that management knows who they are.

Types of Access Control

One type of control involves granting network access to file server(s) containing the drawings. If you are using a local area network or intranet network, the network supervisor can control file access by setting privileges varying from none to read only to full read and write permission. Drawing management software that runs over a network can also be used to augment this rudimentary security by requiring users to check drawings out by registering their name at the time of access to the file. Network access restrictions can be fairly inflexible, however, since they require intervention by the network administrator. That means that the administrator has to be available if access rights need to be changed.

Another type of file access control is possible with or without a network. This system keeps all drawing files on floppy disks—one file per disk. The control of the disks is placed in the hands of the person who would normally control paper drawings. Users are instructed to come to this person to return the changed drawings on the floppy disk at the end of each day, and to pick up their drawings from him or her each morning. Current copies of the drawings are maintained on the network server, or on individual workstation hard disks for use as reference drawings. Some offices collect the disks (after they are copied to the server or main project workstation hard disk) and put them in a fireproof data safe overnight. A security system such as this has the benefit of enforcing the creation of backups of drawing files on removable media.

Who Has the Latest Drawing?

Version control is almost as important as access control. It is critical in the construction drawing phase of a project, because older iterations of drawings need to be available in case the implementation of a change is disputed by a contractor or a client. As I mentioned earlier, long file names give us the ability to document drawing progress. The easiest system to implement is one that uses some form of eight-character basic file name, followed by the initials of the last person to modify the drawing, and the date of the modification. Unfortunately, this means the file name is constantly changing, and if it is a file called as an xref, changing the name removes the latest version of the drawing from the xref "tree."

We used to copy the latest version of the file to a floppy disk that we labeled and dated. We kept five to ten disks in a file box next to each workstation, and each day rotated the oldest version to the front, copying the latest file onto it.

Since we can now use 256-character names for folders, it is just as easy to open Explorer or My Computer, create a new folder (before opening any drawings), and give it a descriptive name and date such as: "Change Order 11.2 revisions 8-14-98". The new folder can hold one or several drawings, depending on how many were modified on that date. Copy the drawings to be modified from their existing folders to the new one, and add some note such as _"old" to the original files' names to take them out of the xref tree. Now start AutoCAD and use the preferences dialog to add the new folder to the xref search path. The drawings will be easy to find because the folder names will display before the file names.

Always Have a Good Backup

> *"He who doth not back up his data shall wander in the wasteland of unemployment, and shall suffer the scorn of the multitudes, and will be beset by troubled sleep and the demons of data loss."*

> CAD Bible, exodusall, v. 10-36

No one needs backup files until there is a loss of the work. Loss of work only becomes a disaster when there are no backup files. There are several degrees of backup that you can implement:

1. Daily copying of drawing files currently modified
2. Weekly backup of all drawing files in project folders (directories)
3. Archive files of drawings issued at completed project phases
4. Backup files of system setup and software configuration files

Backing Up Files onto Removable Disks

I am a firm believer in the use of removable media for drawing backup. The use of removable media is the most flexible type of backup and provides a business with the quickest path to recovery in case of fire or natural disaster, because removable disks can easily be collected and evacuated from a building. Collecting disks, tapes, or disk cartridges should be part of your disaster evacuation plans, and part of your evacuation drills. Backing up files onto removable media should be the responsibility of every CAD drafter. I feel that specific disciplinary actions should be spelled out in the employee manual for failure to back up data as required by an employee's supervisor.

Backing Up Files on a Server

If your computers are networked, you can streamline the file backup process by keeping all current project drawings on a server computer, and installing a removable disk cartridge or tape for performing automatic backups as old files are revised and new ones created. If the building must be evacuated, all the network administrator needs to do is take the removable media cartridges out of the servers and leave.

Principles of Folder (Directory) Structure for Xref Drawings

If All Your Consultants Use Windows 95 or NT

The illustration on the right shows a typical folder structure that I use to track drawing versions, as well as to keep parts of the project in specific locations. The xref structure is very simple, with the drawings in the Leasing Drawings directories attaching the rest room and base building drawings contained in the Restroom and BasePlan directories.

What to Do If Your Consultants Still Use DOS

Your main project directory (folder) can have a long name, such as "5757 Century Blvd". The consultants can name their project directories anything they wish. The folders containing the reference drawings, such as BasePlan, should have names eight characters long (or eight plus a period and three characters, maximum), as shown in this illustration.

The next time we update leasing drawings, we will create a new folder called "Leasing Drawing Updated mm-dd-yy" and then open the

previous versions of the drawings and immediately Save As the same file name, but to the new folder. This creates a new copy of the file in the new folder. Since the xref path to the base building and rest room files haven't changed, all the xrefs will show up as designed.

If we need to make a change to a Base Building file, we follow this procedure:

1. Open the drawing in AutoCAD.
2. Immediately Save As file name_old_mm-dd-yy (for example, 57base5.dwg is saved as 57base5_old_9-12-97, appending the word *old* and the current date).
3. Save As the original file name. This allows us to easily save the drawing periodically as we are working.

Yes, we really do need to change our base building core and shell drawings occasionally, because they were originally copied from paper drawings, and as we are asked to field-verify dimensions, we find discrepancies which must be fixed.

Setting the File Search Path

Now that we have explored the drawing dependency structure and have established a folder and file structure, there is one thing more to do: make sure AutoCAD can find our reference drawing files. Pull down the Tools menu at the top of the screen, and pick Preferences. You will be given a tabbed dialog box as illustrated below.

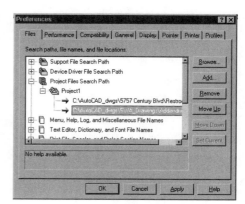

Pick the tab labeled "File System". Pick the **Browse** button next to the window labeled "Support Dirs". You will be shown your current directory tree, and all you need to do is highlight all the subdirectories, right through the last one. Now any intermediate new folders (directories) you insert between the top and bottom folder will automatically be added to AutoCAD's search path, allowing for consistent maintenance of your xref scheme as you develop it.

CHAPTER THREE

DRAWING STRUCTURE AND FORMAT

Contents

As we mentioned in the previous chapters, CAD drawings must be constructed before they can be drawn. This process is *not* analogous to setting up a manual drawing, as you will see. The first thing to be done is to set up a prototype plan drawing with its overall size set to accommodate the largest plan view of the building and/or site and then decide on the internal layer structure of the CAD drawing files. Our Tutorial Building's site is small enough to fit on a standard-size drawing sheet at a scale of 1/8" = 1'-0". Sites larger than normal sheet sizes will require scales small enough to fit them on a single sheet, or they may be plotted on more than one sheet, using match lines. The subject of structure for layers has been an area of controversy for years. It's one of the religious issues of CAD architectural practice. We will shed some light on the issue of naming layers, and search for the truth about how layers should be used.

Setting the Drawing Size

Our Tutorial Building's site is a lopsided trapezoid, 235 feet long at the south side (at the top of the plan), 145 feet in the north-south direction, with the long dimension on the north side being 210 feet. The west end is angled inward from south to north, while the east end is at 90 degrees to the north and south sides.

A simple calculation tells us that a 24" x 36" sheet of paper equals 192' x 288' at a plotted scale of 1/8" = 1'-0". The amount of paper available to our drawing depends on your (or your service bureau's) plotter. Most plotters can plot only within a half inch of the paper's edge (4' at 1/8" scale), which gives us an available area of 184' x 280'. Our overall site, at 145' x 235', will fit on a 24-inch-high sheet leaving us 39 feet on the long sides. At the end of the sheet we will have 45 feet of space remaining.

If we have a 2-inch-wide (16' at 1/8" scale) title block, 29 feet will remain open for site dimensions, which allows 14'-6" on each side of the site. The 39 feet available on the long sides yields 19'-6" at the top and bottom of the site for dimensions. The illustration at right shows how this lays out on the sheet.

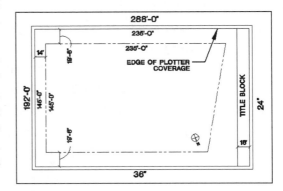

Drawings Are Scaled When Plotted

Changing the scale of the plot at the time the plot is sent to the plotter changes the way the drawing fits on a 36" x 24" sheet. The illustration at right shows the difference in output for a 100' x 50' rectangle, plotted at 1/8" = 1'-0" and at 1/4" = 1'-0".

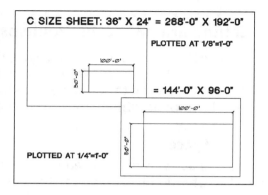

Note that the dimension text and tick marks are scaled up 100% in the 1/4" scale plot. This is also true for any text in the drawing. That is why it is important to design CAD drawings for the scale of the final output.

Always design your drawings so that the size of text and dimension tick marks are appropriate to the **largest** scale you will plot at.

When you plot at smaller scales than your "basic" output size, you will get the effect of a "photo reduction" in text and tick mark size, but they will be scaled in proportion to all the other drawing elements. Most people are generally more comfortable with the visual effects of this type of reduction than they are with seeing 100% increases in the size of symbols and text when drawings are plotted at twice the design scale. Thus, if your building floor plans are almost all plotted in a 1/8" = 1'-0" scale, plotting them in 1/16" = 1'-0" scale will maintain the relationship of the larger scale drawing. Plotting in 1/4" to 1'-0" scale will double the text size and dimension tick size while leaving other drawing elements visually the same as the smaller scale. This looks strange because we are trained to keep text and symbol size the same in manual drafting, regardless of the drawing scale variation.

When drawings are scaled up or down in Model Space (AutoCAD's native default drawing environment), the dimensions change value according to the scale factor. For example, if you have three details on a drawing, and you want to plot one of them at a scale of 3" = 1'-0" and the others at 1/2" = 1'-0", you might opt to set the plot output scale at 1/2" = 1'-0" and scale the other detail up by a factor of 6. Unfortunately, all the values of your associative dimensions in the upscaled drawing would change by the same scale factor. Not only that, all the dimension text and tick marks or arrows are also scaled up by 6 times. If you make the dimensions nonassociative (they don't adjust with the change in the size of the objects they relate to), their text and tick marks' size are scaled up anyway, but their values don't change.

• •
Getting around Scale Problems

AutoCAD provides ways to change the scale of dimensions and text prior to plotting a drawing at a larger scale than it was designed for. Many of these methods have drawbacks, though, in that they work for most but not all dimensions, or text objects.

Paper Space

Paper Space was introduced to help in the scaling problems and is specifically intended to allow drawings of different scales to be plotted on the same "sheet." Drawings in Pspace (AutoCADspeak for Paper Space) retain their associative dimensioning—the dimensions don't change when you scale the drawing up or down. A common practice among many firms is to put the drawings' title block in model space and the drawings in a Paper Space viewport within the title block. This allows a drawing to be plotted at various scales without affecting the scale factor for the title block. It also enables you to plot the drawing on a plot scale of 1 to 1.

The drawing in the Pspace viewport can be dimensioned in Pspace instead of M(odel) Space so that the dimensions scale properly to the plotted scale. Pspace has its own learning curve and other problems that we will address later in the chapter on drawing details.

Use Layers to Separate Scaled Annotation or Other Objects

Larger-scale detail plans are generally not used to duplicate information shown on smaller-scale plans, so the text and symbols for large-scale plans can be on layers different from the same objects on the smaller-scale drawings, allowing the small-scale objects to be turned off in the large-scale versions. Rest room and stair plan drawings are prime candidates for this treatment. This is why we suggest that stair and rest room plan drawings be inserted into plans as reference drawings (xrefs). That way, notes and dimensions can be overlaid by the drawing referencing the basic plan and thus be scaled appropriately to the final output. The chapter on dimensioning will explain how to do this. In a few pages more, we'll explore layers in depth.

Standard American Sheet Sizes in Model Space

Following are some standard sheet dimensions in feet, when plotted at a scale of 1/8" = 1'-0".

We will be setting up our base drawing for the Tutorial Building for a D size sheet in the next section.

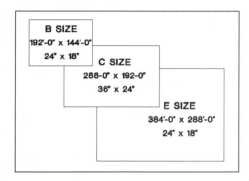

Creating the Base Drawing File

We are going to create a prototype plan drawing "sheet" file, scaled to plot on 24" x 36" paper. Once it is sized properly, we will add layers to it that our plan drawings commonly need, and then save the blank drawing as a template file.

Start AutoCAD. The Start Up dialog will be the first thing to appear on the screen. If you are already running AutoCAD, pick File from the top menu bar, then pick New. Pick the Use A Wizard button, then pick the Advanced Setup bar in the list window.

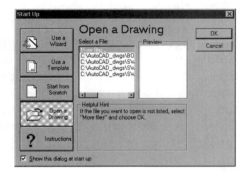

Note: You can suppress the display of the Start Up dialog by unchecking the "Show this dialog on start up" box in the lower left corner of the dialog. AutoCAD LT 95 users need to set the NoStartUpDialog system variable to 1. Use a text editor such as Wordpad to edit the aclt.ini file with AutoCAD LT not running, and save the file.

The next dialog box will be the Units measurement system selector. Pick the Architectural button. For precision, select 1/2 inch. This is probably the most precise measurement a building needs, given the current state of the construction art. You will need greater precision when doing detail drawings, but you can set that when you create your detail files.

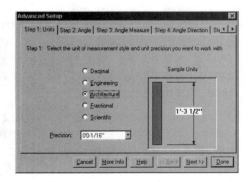

Next pick the forward directional button in the upper right of the dialog frame repeatedly until the rest of the tabs are visible. We are going to leave the settings for angles' units, measurement, and direction set to the normal AutoCAD defaults. Feel free to explore these settings by picking their tabs, but for now pick the "Step 5: Area" tab.

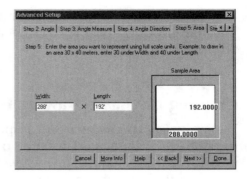

Set the Overall Size for the Drawing

You will now see the Area dialog box. Type 288' in the Width window, and type 192' in the Length window. The illustration shows how these dimensions will relate to the overall size of the drawing. Make sure to type the numbers without spaces, and to include the apostrophe for feet ['] or AutoCAD will give you a 288-inch by 192-inch drawing!

To change a current overall drawing's area, you can use the menu bar at the top of the screen, and enter the new area's values on the command line. From the Format Menu pick Drawing Limits, or type LIMITS on the command line and press [Enter]. The default value of the lower left corner will display on the command line as 0,0. Press the [Enter] key to accept this. You will next be prompted for the upper right corner coordinates. Type 288',192' and hit the [Enter] key.

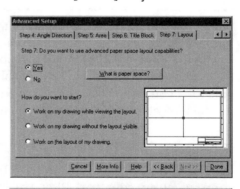

Pick the "Step 7: Layout" button. The illustration shows the dialog box as it will first appear. Pick the "No" button under the "Do you want to use advanced paper space layout capabilities?" question.

This dialog lets you set up Paper Space very easily, but for now, with our emphasis still on the **simplest and easiest** way to work with the CAD system, we will stay with plain Model Space. Press the "Done" button in the lower right corner.

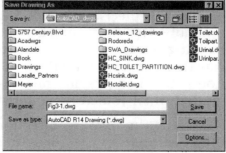

When you have the blank drawing page on the screen, and the Command prompt appears, pick **File** from the menu bar from the menu at the top of the screen. Select **Save As**. The file save dialog window will appear.

Type <u>A-1plan</u> for the new drawing name. Press the **Save** button or the [Enter] key.

You should now have a blank drawing screen with the name "a-1plan.dwg" at the top of the screen, completely set up and ready to draw on. That's all there is to setting up a drawing. The new dialog boxes provided by Release 14's Wizards are a blessing, even for veteran Auto-CAD users. While you can set up drawings from the menu bar and command line, the Advanced Wizard gives you all the control of the old methods, with much more clarity. Now for the layers.

• • • • • • • • • • • • • •

What Are Layers?

Layers in an AutoCAD drawing are somewhat like imaginary underlay and overlay sheets. One layer may have only the walls, another the stair drawings, etc., and they have line weights (associated with colors) and line types (solid lines, dashed lines, center lines, etc.), just like paper overlay drawings.

The big difference between manual and CAD drawings is that in CAD drawings layers are not really separate drawings (sheets) but are part of a single, unified drawing, contained in a single file in the computer. Layers add dimension to CAD overlay/underlay reference drawings (xrefs), allowing for the control of visibility of information as **drawing files** (sheets) are placed on top of one another. For example, I can use my reflected ceiling plan to call the floor plan as a reference drawing, and I can turn off the floor plan's doors and door swing arcs so they are not visible on the reflected ceiling plan drawing.

This diagram of a simple building plan shows the **concept** of how entities in a drawing are structured by layer.

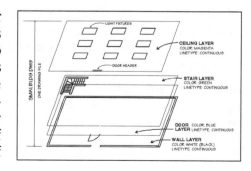

In the diagram, you will notice that the layer color and line type are listed with the layer's name. **Layer colors** are used by AutoCAD to assign pen sizes to particular line weights when plotting the drawing. Line weights are not graphically shown on the AutoCAD display screen. This has always made it hard for CAD drafters to visualize the final output of the drawing, but it has the added benefit of being **plotter independent**. I can draw a building with layer colors mapped to my plotter, and my consulting engineers can take the same drawing and remap the layer colors to their plotter without changing any of my layer colors or layer organization. When the engineer and I exchange electronic versions of the drawing files, my layer/line weight map will prevail on my end, and theirs on their drawings, because Auto-CAD keeps that information separate from the actual drawing file. Unfortunately, we can't have it both ways; however, I think there are definite advantages to not locking line weights into drawings used by several disciplines.

Line Types Commonly Used in Architectural Drawings

ACAD Name	*Generic Name*
Continuous	Solid
Hidden/Dashed	Dashed/Dotted
Center	Center Line
Phantom	Property/Boundary

Here's what they look like:

Release 14 provides many more types of lines than previous versions of AutoCAD, many specifically tailored to civil engineering and mapping. It also provides an easy method of creating your own line types, but I recommend that you avoid too many custom line types, especially if you share drawings with consultants who are using older versions of the program.

Lines are referred to as having "weight" rather than "thickness" in ACADspeak, because the term *thickness* refers to the 3D height of the line above the "floor" plane of the drawing. All AutoCAD lines are 3D objects; their default thickness is always zero unless the drawing is set up to override that default setting.

Line types are scaled independently of other drawing objects such as circles and polygons. The line type scale (LTSCALE) system variable controls how many dashes per foot the line may have (for the dashed line type).

Note: You can change the line type scale at any time in a drawing just by typing LTSCALE on the command line and pressing [Enter], then typing in a new numeric value. Higher values result in larger scale spaces and dashes, lower numbers in smaller differences between the line's elements.

Switches and Temperature Controls

Layers can be turned on and off by dropping down the Layer Control dialog window on the Object Properties tool bar. They can also be Frozen and Thawed. This is easily accomplished just by picking the layers and picking the lightbulb symbol at the left of the layer name in the Layer Control window, shown on the next page.

The Layer & Linetype Properties dialog window is accessed by picking the [] Layer Control button at the top screen tool bar. We will use that button to set up layers for the drawing shortly, but first let's talk about how layers are manipulated.

Turning a layer **off** or **freezing** it makes all the entities on it invisible on the screen, and prevents them from being picked by accident.

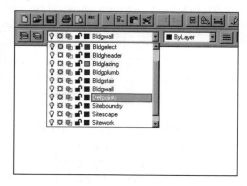

Layers that are turned **off** continue to be invisibly redrawn whenever you ZOOM in and out or PAN the drawing view. This can make redraws slower than if you froze the layer instead. When the layers are turned back **on**, AutoCAD redraws the screen, which is always faster than regenerating it (see next paragraph). To turn layers **on**, pick the dark lightbulb on the drop-down layer list on the tool bar. To turn layers **off**, pick the "lit" bulb.

Freezing a layer prevents it from being redrawn during Pans and Zooms but forces a regeneration of the entire drawing when it is **thawed**. Using **freezing** instead of the **on/off** switches is always a matter of experimentation and personal preference. To freeze a layer, pick the sun on the drop-down layer control window. To thaw it, pick the blue sun as shown in the illustration.

Note: Blocks (objects inserted using the Insert Block Tool or INSERT command) will remain visible when the layer they were inserted on is turned **off**. Layers with inserted blocks must be **frozen** to render the blocks invisible. This is also true for blocks on xref drawings. If you are working in a viewport, in either paper or model space, picking the sun freezes/thaws all layers in all viewports. Pick the sun symbol with the rectangle to freeze layers only in the active viewport.

Locking a layer prevents objects on it from being accidentally selected when you are drawing. To lock or unlock a layer, pick the padlock symbol next to the layer name on the drop-down layer control window.

Making a layer current (the one you draw on) is much easier in Release 14 than before. You can drop down the layer list and pick the layer name, or you can pick the "**change to**" layer button (🖹) on the tool bar, then pick an object on the layer you want to work on. This has the benefit of your not having to know what layer the object is on; you can just go there and keep drawing.

WARNING: As with all good powers, this feature has a dark side. If the object you pick was drawn on the wrong layer, any new drawing you do will just compound the problem. Drawing on the wrong layer can be a severe time waster. Moving all the objects created incorrectly to the desired layer can only be accomplished by tedious manual selection of each and every object.

Picking the 🖹 Layer button on the tool bar brings up the Layer & Linetype Properties dialog box. This dialog tool lets you change a layer or group of layers' colors, line types, visibility in the drawing, and visibility in the drop-down layer list. We will use it in the next sections to create the layers for our plan drawing, but first we need to decide just what layers a building floor plan really needs.

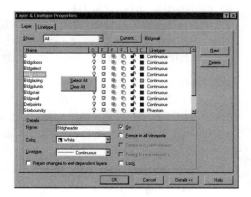

Designing a Layer Structure for a Drawing

From its beginnings, AutoCAD has provided us with the blessing and curse of unlimited layers available in each drawing file.

Elaborate layering systems were once needed because they were the only way to separate drawing information in older versions of AutoCAD, before the advent of reference (XREF) drawings. Now we can use reference drawings to overlay and underlay separate drawing files, as explained in Chapter 2.

We also do *not* need complex layers for multiple line weights and line types in most architectural plan drawings. Manual drafters make do with only three or four line weights per drawing. CAD drafters should do the same, and should also use simple layer schemes.

> *The goal of a layering system should be to streamline the process of creating 2D construction documents, and should never add unnecessary complexity to the process.*
>
> *The biggest failures in CAD drafting happen when the CAD drawing's layer structure is poorly designed.*

A badly designed layer system will be unenforceable; project team members will create chaos by making well-meaning modifications just to do their work. As the layer structure falls apart, productivity plummets, because older generation drawings will need to be "fixed" over and over again.

Setting Up Floor Plan Layers

The first step in designing the layer structure is to **analyze a set of existing drawings** similar to the one you are about to create. Get a set of completed plans from a similar project. Compare the floor plan to the other plans, and color all parts of the floor plan that do **not** appear on the other plan drawings. Use different colors for different drawing elements such as stairs, notes, and doors.

Now answer these questions for all the items you have identified as unique to the floor plan:

1. Will I ever need to make this item invisible while drawing or plotting?
2. Does this item need to be plotted in a line weight other than medium?
3. Do I need to see this part of the drawing in a different, distinct color on the monitor screen?

If the answer to any of the three questions is no, do not create a new layer for that part of the drawing.

We don't expect you to be able to answer any of these questions now. The next section will show the layer and color/line type structure for our Tutorial Building. Remember the questions as you work with the building drawing, and you will discover how to answer them on your own projects.

Basic Rules for Layer Structure

1. Limit the number of layers **per drawing file** to no more than twenty-two, which is equal to two screens of the Layer & Linetype Window. It is time wasting to scroll through long layer lists, and it is doubtful you will need more than three line weights spread among twenty types of objects to draw architectural plans and elevations.

   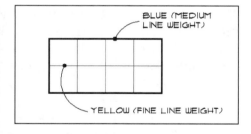

   If you need different line weights in objects that should be on one layer, you can change the line color (plotted weight) independently of the assigned layer color. For example: We have a 2' x 4' fluorescent light fixture with louvers on layer Clgfixt. We want the louvers to plot lighter than the fixture outline, so we change their color to yellow, which we have decided will be our fine pen weight color.

2. Plan your layer structure on paper before putting it in a CAD drawing. Mark up design plans and any other sketches with layer names you plan to use, keyed to the parts of the drawing they will represent.

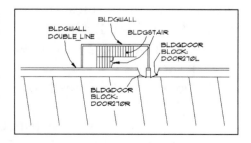

For our building drawing, we have started to do just that. On the right is a zoomed-in view of an exit stair area and its layer structure:

Layers are listed first, then the drawing element name (Circle, Arc, Double_Line), with the line type listed last. **All line types are continuous unless noted otherwise.** The stairs are separated on their own layer, as are the doors. This will let us turn them off (making them invisible) in the Site, Reflected Ceiling, and Roof plans. The doors are drawn once, then turned into entities called Blocks, which can be inserted in the drawing many times.

Layers for the rest rooms provide a separation between the toilet stalls and the plumbing equipment so that the plumbing plan can show the equipment without the stalls getting in the way of dimensions. Detail plans for stall dimensions and construction can be extracted by layer, without taking the plumbing fixtures along.

Plumbing fixtures and stalls will be created as separate, inserted drawings, to enable us to reuse them in other buildings, and to aid in planning future rest rooms.

Exit stairs and doors are on the same layer as other stairs since they are part of all plans except the roof and foundation drawings.

As you mark up plans and other drawings for layer assignment, make a list of each layer name you create, to avoid creating duplicate or confusingly similar names.

Layer Name Structure

Names for layers should be highly mnemonic (they should be easy to interpret and memorize). The goal of naming layers is to make the names easily readable at a glance in the current layer display window on the Object Properties tool bar. **Avoid codes and abbreviations that must be memorized.**

The Main Constraints in Naming Layers

Layer names are limited to a maximum of thirty-one characters.

Names may contain only letters, numbers, the hyphen (-), underscore (_), and the dollar sign ($).

Only the first twenty or so characters of any layer name are visible in the Current Layer Window in the Object Properties tool bar, and the dropdown layer window. This varies depending on your screen resolution and the font you select in AutoCAD's Preferences.

Layer names should be readable at a glance. Use all thirty-one characters if necessary. A layer name whose meaning has to be guessed by someone unfamiliar with your system will only create ill will among your consultants and even your clients. Why name a layer A-ELVOT (as in the proposed 1996 AIA guidelines) instead of BLDG_OUTLINE, which is what A-ELVOT is supposed to stand for?

What about the AIA Layer System?

The current AIA layer-naming system was designed in an era when reference drawings were not available in AutoCAD. In those early drawings, all objects belonging to a floor plan, such as ceilings, dimensions, and electrical equipment, were drawn in the same file. Keeping these virtual drawings separated required large numbers of layers. The AIA system was an attempt to decode layering in CAD files, where earlier schemes relied on numbers (sequential 1, 2, 3, etc. or the C.S.I. materials codes: 1000 for site work, 2000 for metals, etc.). It also helped identify which discipline each group of objects in the drawing belonged to by providing an **originator code** letter such as A for Architectural, C for Civil, and so on. This was a major advance at the time, but times change. Unfortunately, the draft revision proposed by the AIA for updating the system makes little progress over the past one, solving none of the current system's shortcomings.

The AIA system adds hyphens, which use up space in the Current Layer display window without making any contribution to the layer's legibility. It also restricts each section of the layer name to four characters, which limits the mnemonic quality of the name as well as its ease of recognition.

The system also identifies each layer by a discipline code: A- for Architectural, P- for plumbing, E- for Electrical drawings, and so on. These prefixes at best are redundant, and at worst further shorten the displayable part of the name without adding meaning. I maintain they are redundant because with a good Reference Drawing system, only electrical layers go on the Electrical Plan drawing file, only plumbing layers go on the Plumbing Plan drawing file, etc. When you open an Electrical drawing file, you obviously know which drawing you are working with! Likewise, the building's walls will be in the Architectural Floor Plan file, not the Electrical Engineer's overlay drawing. Abandon the All-In-One Drawing model and you abandon the need for discipline codes in layers.

The following table shows the layer names we will be using, contrasted to their AIA equivalents:

Our Layer Name	*AIA Layer Name*
Bldgdoor	A-door-bldg
Interior_doors	A-door-intr
Bldgwall	A-wall-bldg
Partition	A-part-intr

Layer names that are easy to read at a glance are very important because **one of the easiest errors to make in CAD drafting is drawing on the wrong layer.** It results from not noticing the difference between layers of the same color. **The layer name is your only clue to where you are.** I have seen hours of time wasted because someone drew walls on what they thought was the wall layer, only to find out later that it was the **text layer** of the same color. None of us is immune from this mistake. The best we can do is arm ourselves with layer names that will show us at a glance our true location in the drawing structure.

● ● ● ● ● ● ● ● ● ● ● ● ● ● ● ● ● ●
Layer Color Structure

Mapping colors to layers has only one practical constraint: Colors should be selected for their contrast with the screen background. High value and/or tint contrast will ease eyestrain, which should be your primary consideration. Unfortunately, it is very difficult to see the difference between shades of color, making the use of any more than the first eight colors used by default in AutoCAD a futile exercise.

Color choice may be limited by a number of factors, such as the color capability of your display system, the resolution of your monitor, and whether or not your consultants and team members are using your preferred background color. It is important for you to choose colors that you are comfortable with personally, but imposing your personal tastes on team members and consultants is never wise.

If your firm already has a standard pen weight relationship to specific colors, use it. Otherwise, select pen weights at the same time you select colors to insure that all entities drawn on a given layer will be plotted with the correct line weight.

Remember, layers can contain objects with different colors if necessary for plotting. Use this tactic sparingly, as it destroys the only other visual clue that tells you which layer you are drawing on.

Color, Contrast, and Ergonomics

The human visual system perceives contrast between light and dark before it sees differences in color. The selection of the screen background color directly affects the visibility and contrast of certain object colors. Black and white screen backgrounds provide the highest contrast for some colors, but they may be a poor choice for ergonomic and practical reasons.

Ergonomic Problems

Veiling reflections of light-colored background objects such as walls are more prevalent in a black background and can be a major source of eye-strain. Looking back and forth between paper drawings and a black screen over a period of hours is more stressful than using a screen background color that is closer to that of paper.

Practical Problems

Both white and black screen backgrounds do not allow some colors to be easily seen. Light value colors such as green, yellow, and cyan do not show up well on a white background. Yellow is practically useless. On a black screen background, medium gray (color 8) and blue (color 5) are very difficult to see, as are violets and reds in the darker ranges. This effectively limits your color choices available for layers to at most six distinct colors.

I have found that the best choice (for me) is to set the screen background to a medium gray. This displays lighter colors with sufficient contrast and allows me to use black lines. Dark colors such as blue are also very easy to see.

Remember: Layer names that are easy to read and understand in the Current Window 20 character display are the best defense against drawing on the wrong layer.

Creating Layers for the a-1plan Drawing

Pick the 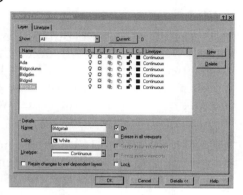Layer button on the tool bar, to open the Layer & Linetype Properties dialog window, as shown below.

Add Layers to the Drawing

Click on the Details>> button at the bottom of the dialog to expand the window if it does not look like the illustration. (Note: Once Details is picked it changes to read Details<< as shown in the illustration.) Pick the **New** button, and you will immediately see a gray edit field rectangle with the layer name "Layer1" in it. Type the Layer List as shown in the illustration.

Type a name, then a comma, followed by the next name. Do not insert spaces between the commas or names. Do not press [Enter], or you will have to repick the New button. Typing commas at the end of the name moves you to the next line, ready to enter the next name (it's an ACAD "feature"—you really don't want to know why).

Ada	Bldgplumbing
Bldgrid	Bldgstairs
Bldgcolumn	Bldgwalls
Bldgstairs	Intequip
Bldgdim	Intsoffit
Bldgdoors	Intwalls
Bldgelect	Siteboundary
Bldgheader	Sitenote
Bldglazing	Sitework
Bldgnote	

Loading Line Types into the Drawing

We have not previously loaded any line types into our drawing, so we must do that now. The easiest thing to do is load all the AutoCAD line types, then later get rid of the ones we don't need.

Press the **Linetype** tab on the Layer & Line-type Properties dialog window. Press the **L**oad button.

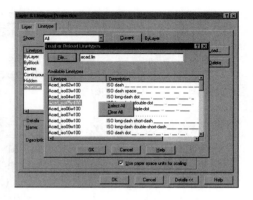

Press the right mouse button to pop up the Select All and Clear All box (shown in the illustration). Pick **Select All**, and all the line types in the list window will be shaded. Pick the **OK** button. The line types will now load themselves into the drawing.

To load only selected line types, hold the [Ctrl] key down and pick the types you want, then pick **OK**. You can also use the [Shift] key just like in Windows 95 Explorer to select a contiguous group of line types.

Other Layer List Controls and Tools

Changing a Layer's Line Type

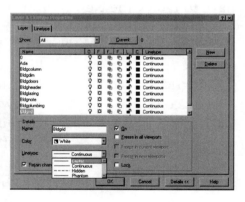

1. Pick the **Layer** tab to open the Layer dialog window again.
2. Pick the Bldgrid layer.
3. In the Details section, pick the Linetype drop-down list, then select the Center line type in the Linetype list window.

Repeat this process to give the Siteboundary layer the Phantom line type.

Repeat the process again to give the Intsoffit layer the Hidden line type.

You can change the line type for many layers at once by first picking their names while holding the [Ctrl] or [Shift] key down, then dropping down the list and picking the line type to assign to all the selected layers.

Now we're ready to change layer colors so they will be easier to distinguish on the screen.

Changing a Layer's Color

1. Pick the layer or several layers you want to change to a new color. In our example, we are going to pick the Bldglazing layer.
2. Pick the **Details>>** button to expand the dialog window, if necessary.
3. Pick the drop-down color list, and select a color. We will select cyan for the glazing layer.

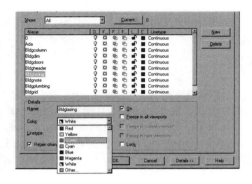

In the Layer List window, you will see that the color has now been changed for all the selected layers.

Using Filters to Edit the Layer List

In the upper left of the dialog is the Show list window, or Filter drop-down list, as AutoCAD likes to call it.

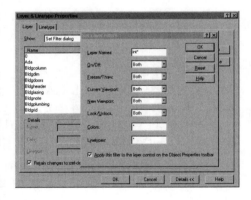

Setting filters for layers is a powerful way to navigate quickly through a long, tangled layer structure. To activate the Filters dialog box, pick the button on the Show window's right, then pick the Set Filter dialog... item at the bottom of the drop-down list.

In the example above, we have typed in **"int*"** in the Layer Names box. This automatically filters out all layers whose names do not begin with **int**.

It is important to use the asterisk (*) to represent the rest of the layer name. If you omit it, AutoCAD will search only for a layer named **int**. Putting the asterisk in front (***int**) searches for all layers **ending** in **int**. With our filter set, we activate it by picking the **OK** button.

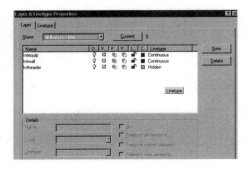

The Show window now reads "All that pass filter," and all the layers that start with **int** are displayed, as shown in the illustration. We can now use the Shift key and the mouse or arrow keys to highlight all our filtered layers and then perform any global modification we want, such as changing their line types or colors, turning them off or on, or freezing or thawing them.

Alternatively, we can modify them separately. The filter function saves time in scrolling through lists of layers—it can be a real productivity booster. Once you have set a filter, it will remain active until you disable it. To cancel a filter, open the Set Layer Filters dialog box again and type an asterisk (*) in the Layer Name window, then pick **OK**.

We can filter layers by color as well as line type simply by typing the color or line type name in the appropriate windows. Setting the Freeze/Thaw window to Freeze or Thaw performs that operation automatically on the filtered layers. The Lock/Unlock window works the same way. The Both setting (the default) leaves the layers unaffected.

The Current button is used to make a selected layer the active layer (the one being drawn on). This button cannot be used when more than one layer is selected.

Deleting Layers

For the first time in AutoCAD, it is possible to delete layers by selecting them as described above, or individually, and pressing the Delete button. Previous versions of AutoCAD required that the PURGE command be used to accomplish this.

Layers that have a block definition referencing them or that have a reference drawing inserted in them or are associated with a text style cannot be deleted until the referencing objects are PURGEd from the drawing, and in the case of the reference drawing, detached.

Renaming a Layer

Renaming a layer is easy. Just highlight its name and type a new one in its place.

Turning Layers On and Off; Freezing and Thawing

This control turns highlighted (selected) layers on and off.

The sun symbol freezes and thaws a layer or group of highlighted layers.

The shaded sun with the rectangle behind it controls the freezing and thawing of layers in the *current* viewport, in either Model or Paper Space.

The shaded sun with the rectangle in front of it controls the freezing and thawing of layers in all viewports.

Sorting the Layer Control Display

AutoCAD displays all layer names in alphabetical order, in the Layer List window, unless they start with a number, in which case it displays them in numerical order, followed by alphabetical order. Clicking on the Name header bar in the list window sorts all layer names alphabetically; clicking on it again reverses the alphabetical sort order.

Now you can see how keeping the layer quantity to twenty per drawing allows rapid movement through the Layers & Linetype Properties list. If you use more than twenty layers, you will be forced to use the scroll bar to access the middle areas of the layer list every time you perform routine operations. Setting a filter to show <u>All in use</u> will help if your drawing contains many unused layers.

We have finished our initial layer setup. We are next going to save our drawing and its layers as a template drawing (a drawing with the .dwt extension on its file name).

• •

Saving the a-1plan Drawing as a Template

Pick File on the tool bar and pick Save As from the drop-down file menu. You will now see a Windows 95-type File Save dialog unique to Release 14. The file name "a-1plan" should already be displayed in the File Name window at the bottom of the dialog. Drop down the Files of Type list as shown in the illustration and select Template.

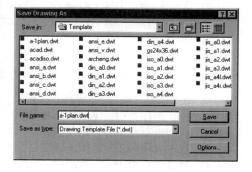

In the pop-up dialog, type a brief description such as <u>Architectural Plan-36x24 at 1/8" = 1'-0"</u>, which will display whenever template drawings are selected in the File Open dialog. Press the Save button to save the file.

Template drawings are roughly equivalent to prototype drawings in Release 13. They are also available in AutoCAD LT 97 but are a new addition since Release 13. I recommend that template drawings be kept in the default Template folder provided by AutoCAD for easy location. If you are using a network, set up the Template folder on a server where everyone has access to it.

Using the Object Properties Tool Bar

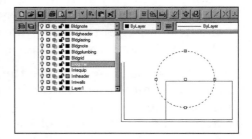

In the upper left side of the Object Properties tool bar is the pull-down Layer List, which works just like the Layer List in the Layer & Linetype Properties dialog window. Picking the arrow to the right of the Current Layer window will drop down a scrollable list of layers, displaying ten or more names at a time.

Using the Layer List

The current layer is always highlighted. Picking any of the other layer names immediately switches the current layer to the one picked, making changing current layers on the fly a two-mouse pick operation. As we mentioned before, you can also use the 🖼 Make Current Layer button to change the current layer to the same one as that of any selected object. Still more capability is offered: Pick any object, then drop down the layer list. Pick a layer to move the selected object to, and then press [Esc] or [Ctrl]+[C], and the object will be moved to the selected layer.

Changing an Object's Color with the Color List

The Current Color window can be dropped down to select a new Current Color to override the Bylayer setting. All objects drawn after a new color is selected will be on the current layer, but will have the selected color. Any change made with these controls does not affect anything drawn before the change; only parts of the drawing drawn after the change will reflect the new settings.

The Current Color window can also be used to change the color of a previously selected object or group of objects, as we described for the Current Layer list.

Picking the Linetype button calls up the Layer & Linetype Properties dialog with the Linetype tab selected.

In the next section we will set up a title block drawing, which will showcase some of Release 14's ability to handle bit map graphics and True Type fonts.

CHAPTER FOUR

TITLE BLOCK DRAWING CONSTRUCTION

Contents

Title blocks are not something most architectural drafters draw every day. Once set up, they are rarely modified (our firm, Partners By Design, is an exception). However, Release 14 offers some compelling new graphics support that may greatly affect how your company logo can be used in your title block drawings, and you may find yourself redesigning your drawings after you upgrade to the new software. Of course, this new graphics integration can also introduce incompatibilities with consultants who are using Release 13 or 12, which don't support hybrid graphic documents or the new filled area data type of Release 14. Don't throw away the old title block file just yet. Some of your projects may still require it!

Title Block Drawing Types

The prevailing wisdom in title block use is that it is best to draw one title block and insert it in a Paper Space view. The drawing is then inserted into another Pspace viewport, where it can be scaled independently from the title block and border. This remains the base sheet setup for every drawing file. To see how this works, bring up the Siteplan.dwg drawing from AutoCAD's sample files (a modified version of this drawing file is included on this book's companion disk in case you don't have access to the sample files).

There are several problems with this approach. First, you are stuck with the title block as part of your drawing, from the beginning and forever. This is fine for real paper drawings because we are used to identifying our drawings with the information on the title block. It can be a major pain in CAD drawings because we identify them by file name, in Windows Explorer. It is helpful to see confirmation of the name on the drawing screen, but not essential, and title blocks contain lots of text and graphics that are very slow to redraw, and even slower to regenerate. A major problem with Paper Space is that all viewports and the Paper Space objects redraw whenever you use the REDRAW command. One of the companies I consult for has a fairly ordinary title block that takes ten seconds to redraw on its fastest machine, and over twenty seconds on its slowest. Redraws happen with every ZOOM and PAN outside the Paper Space viewports, and the pause caused by the title block is very irritating. A major improvement over Release 13 is that Paper Space viewports do not regenerate with every PAN and ZOOM anymore, and Release 14 allows us to XREF the title block drawing into Paper Space, and then to temporarily detach it while working on the drawing using the UNLOAD command. This preserves the path data of the xref, while giving us a major improvement in drawing speed.

Second, you are not relieved from setting up separate dimension styles for each scale factor you want to use in a drawing. If your Paper Space drawing contains a plan at 1/8" = 1'-0" scale, and an elevation at 1/4" = 1'-0", you will need a Dimension Style for each of them, with the dimension scale for the plan set at twice the size of the elevation's. The alternative, advocated by many AutoCAD authorities, is to do your dimensioning in Paper Space, using Paper Space scaling. This can work, except that when you bring up a model space view of our hypothetical plan, the dimensions would vanish, because they exist only in the Paper Space view.

Third, if you wanted to edit our hypothetical elevation and plan drawings while in Paper Space, they would both need to be contained in a single file. This is not a bad approach if your files are the size of the sample Siteplan.dwg, but a large sheet of details can consume several megabytes of disk space, making it difficult to share with consultants. Large files also provide lots of coffee-break time while they redraw and regen. They also prevent anyone else but the current user from working on parts of them, as in the case of our elevation/plan example: If I am updating the rest rooms, my team members are unable to work independently on the elevation(s) while I'm using the file. During the phases of the project when several people must be working simultaneously on related parts of the drawings, using the **single file holds everything** method can seriously impede drawing production.

Fourth, the use of Paper Space just introduces another layer of complexity to the CAD drawing process. Added complexity means more training time, and more opportunity for error. This is too great a price to pay for consistent plots of the title block.

I advocate setting up title block drawings for each scale commonly used in your work. These drawings are inserted as xrefs when you plot but for the most part are left out of the drawing. This makes work on large drawings much faster, and the drawing name and number are still part of the model space drawing, not the title block, so you see visual confirmation of the drawing file's identity on the screen.

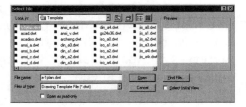

Setting Up the Title Block Drawing File

Start AutoCAD and select Use a Template from the Drawing Start Up dialog, or select the Open File button on the tool bar, and select the Template file type in the Open File dialog. Pick the a-1plan.dwt file we created earlier.

 1. Drop down the Layer & Linetype Properties window.

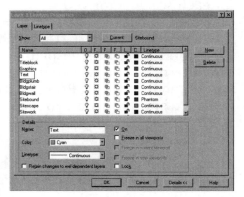

 2. Release 14 allows you to type the name directly in the Layer List window, or you can use the Name window below the list. Edit the first layer on the list to change its name to Titleblock. Change the second layer's name to Text, and change the third layer's name to Graphics. Edit the fourth layer's name to Sheet_Border.

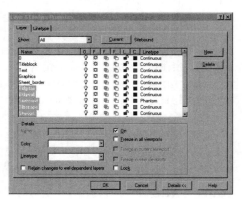

 3. Select the fifth name on the list, and holding down the shift key, scroll to the last layer name and pick it to select the entire list of remaining names.

4. Pick the Delete button to the left of the Layer List window.

Go ahead and assign the three layers a color of your choice. All line types should remain continuous. Pick the OK button to close the Layer & Linetype Properties window.

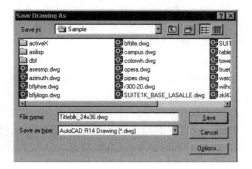

Now select the Save As option on the File dropdown menu, and save the file as "Titleblk_36x24.dwg" with the file type set to Drawing File. The file name displayed in the status bar at the top of the screen should now show the new file name.

Title Block Design and Other Religious Issues

I'm sure the title block and other sheet design elements used by your firm are the ultimate in design function and expression. I am not suggesting that you change a line of your current office standard! In my practice, we change our title block several times a year because of the diversity of our practice. We also sometimes try to incorporate the corporate logo of our clients into the title block, which means shuffling things around to make space for it. Release 14 makes this process infinitely easier than ever before, as we will show in the following tutorial.

Borders: In or Out?

Drawing borders can be problematic in CAD drawings because plotters other than the one for which the border was set up may not be able to plot as near the paper edge and will fail to plot one or more sides of the border. This situation may create headaches for your client or consultants. On the other hand, borders do establish a fairly consistent way to constrain a drawing so that it doesn't have parts falling outside the plotter's range. One reason we are forced to change our CAD title block drawings is that we want to make sure everything works on other parties' plotters on each project. Personally, I've given up on borders. They are just one more headache I don't need when I've got enough to worry about in getting the drawings out the door.

In manual drafting, the title block is printed on a sheet of paper and therefore never varies. In CAD drafting, this is almost never true in practice, because of the demands different plotters make on the final format of the drawings.

Enclosed or Open Blocks?

It's up to you. I have selected an enclosed block for the tutorial, simply because the methods for creating it also apply to borders. They also somewhat apply to "stacks of blocks" type designs. The first design decision, then, is to determine the width of the title block border line, plotted at 1/8" = 1'-0" scale. I generally use 8", and that's what we will do here. The overall width of the block will be 24'-0" to the center of the border line, and it will be spaced 6'-0" in from the edges of the sheet, to make sure that our plotter will not cut it off 1/2" from the edge margin.

The first step is to draw the sheet outline so we can be visually oriented to the design.

Drawing the Sheet Border

Change the current layer to Sheet_Border by using the drop-down layer list on the tool bar.

Zoom out to the drawing's limits by picking the ZOOM All button on the tool bar.

 Type <u>PLINEWID</u> on the command line and press [Enter]. Type <u>8</u> + [Enter]. PLINEWID is the system variable that controls the default width for Polylines. It cannot be set on the fly when using the RECTANGLE command as it can be when using the PLINE command for drawing individual Polylines.

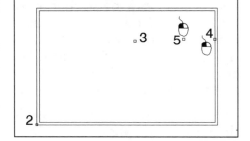

    1. Pick the RECTANGLE tool.

    2. Type <u>0,0</u>+[Enter], for the starting point of the rectangle.

    3. Type <u>288',192'</u> and [Enter] for the ending point. We have just used a polyline to create our sheet outline.

    4. Pick the OFFSET tool from the Modify tool box.

    5. Type <u>6'</u> and [Enter].

    6. Pick the sheet outline.

    7. Pick a point inside the sheet.

We now have a second Polyline 6'-0" inside the first. We will now change its shape to the 24-foot-wide title block we want. First though, we are going to set the sheet border line's width to 0 by exploding it.

Drawing the Enclosed Block Title Boundary

    1. Pick the EXPLODE tool from the Modify tool box.

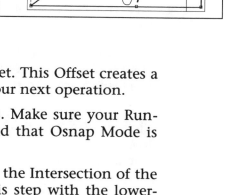

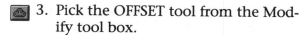 2. Pick the sheet outline. Press the mouse's Enter button.

3. Pick the OFFSET tool from the Modify tool box.

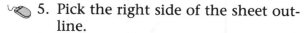 4. Type <u>30'</u> followed by [Enter] (24' for the width of the title block plus 6' for the margin offset).

5. Pick the right side of the sheet outline.

6. Pick a point to the left, inside the sheet. This Offset creates a quick and dirty construction line for our next operation.

7. Pick the Polyline to turn on its Grips. Make sure your Running Osnap is set to Intersection, and that Osnap Mode is on.

8. Pick the upper left Grip and drag it to the Intersection of the construction and Polyline. Repeat this step with the lower-left-corner Grip.

9. Press the CANCEL key(s) twice to exit Grip mode.

10. Erase the construction line (you can use a Window to select it to avoid a difficult pick or zooming in and out). We now have our basic title block boundary. The next step will be to move it to the Titleblock layer, then to divide it up. Pick the boundary line to highlight it.

11. Pick the Properties tool (DDCH-PROP) on the Object Properties tool bar. The Modify Polyline dialog box will appear.

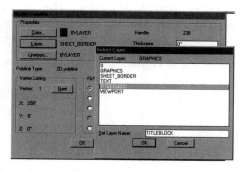

12. Pick the Layer button. The Layer Control dialog should now be visible. Pick the Titleblock layer. Pick OK.

13. Pick OK on the Change Properties dialog box, and the boundary will now be on the correct layer.

Create the Drawing Data Box

Even with the boundary on the correct layer we are still not ready to draw, because a quick glance at the current layer window in the tool bar will show that we are still on the Sheet_Border layer. Pick the ▤ Make Object's Layer Current button and pick Titleblock to make its layer (Titleblock) the active one.

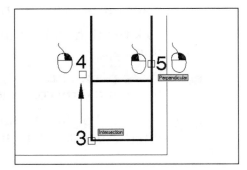

1. Pick the PLINE tool from the Draw tool box.

2. Pick the TRACKING tool from the customized Draw tool box, or from the Osnap area on the Standard tool bar.

3. Pick the Intersection of one of the lower boundary corners as the first tracking point.

4. Drag the pointer up (90 degrees) and type <u>24'</u>, followed by [Enter]. Press the mouse's Enter button or the [Enter] key to end Tracking.

5. Pick the Perpendicular Osnap button on the tool bar. Pick the opposite vertical boundary line, then press the mouse's Enter button to end the Pline command.

Note that we did not need to specify the width of the Polyline this time, because AutoCAD keeps the last width used as the default for all future uses of the command.

Creating a Lighter Line for the Lower Side of the Box

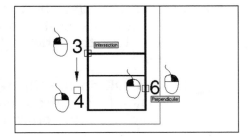

1. Press the Mouse's Enter button to recall the Pline command.

2. Pick the TRACKING tool from the customized Draw tool box or from the Osnap area on the Standard tool bar.

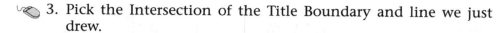

3. Pick the Intersection of the Title Boundary and line we just drew.

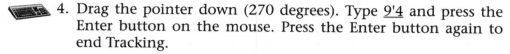

4. Drag the pointer down (270 degrees). Type 9'4 and press the Enter button on the mouse. Press the Enter button again to end Tracking.

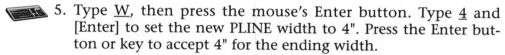

5. Type W, then press the mouse's Enter button. Type 4 and [Enter] to set the new PLINE width to 4". Press the Enter button or key to accept 4" for the ending width.

6. Pick the Perpendicular Osnap button on the tool bar. Drag the Pline to the other side of the Title Boundary and pick the vertical boundary Pline. Press the mouse's Enter button to end the Pline command.

Zoom in to the view shown in the next illustration. Change to the TEXT layer, using the drop-down layer list on the Object Properties tool bar.

Placing Paragraph Text in the Drawing Data Box

1. Pick the Paragraph Text tool (MTEXT) on the Draw tool box.

2. Pick a starting point about 2 feet inside the Data Box, and drag the text window down to the lower right.

3. Pick the lower right corner of the text area. This will automatically open the Multiline Text Editor window shown in the next illustration.

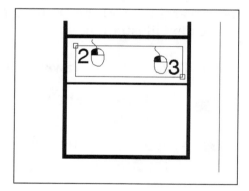

Using the Multiline Text Editor

1. Select the font you want to use from the drop-down list on the far left, just above the editing window. We have picked a Helvetica type font called helv01.shx (provided with Release 14), but you can obtain sim-

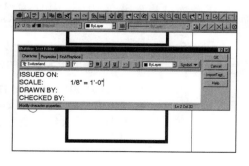

ilar results by selecting one of the Swiss fonts supplied by AutoCAD, such as SWISS 721 BT. The double T to the left of a font name identifies it as a True Type font.

 2. Set the font height to 1 foot: Pick the size list (to the right of the font name window), and type 1' followed by [Enter]. Do not try to type with the size list dropped down, as AutoCAD will not accept any keyboard input when the list is deployed. The text height should now display as shown in the illustration on the previous page.

 3. Pick the text window to make the text cursor active, and start typing the text shown in the illustration. Press the [Enter] key at the end of each line. Use the space bar to space over to the location of 1/8" = 1'-0".

 4. Just after the line "CHECKED BY:", create a blank line by pressing [Enter] without typing anything else on the line. Finish typing DRAWING NO. and pick the OK button.

Your text should now be properly positioned, with DRAWING NO. located at the top of the Drawing Number box. Zoom to the view shown in the next illustration.

Create Boxes for the Drawing Name and Title

 1. Pick the OFFSET tool on the Modify tool box.

 2. Type 24' and press the [Enter] key.

 3. Pick the top line of the drawing data box.

 4. Pick a point above the line, as shown in the illustration.

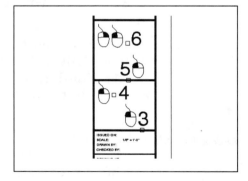

5. A new line will appear 24 feet above the data box. Pick the new line.

6. Pick a point above the new line and press the Enter button on the mouse to end the Offset command.

Create the Revision Block Space

 1. Pick the OFFSET tool on the Modify tool box.

 2. Type <u>48'</u> and press the [Enter] key.

 3. Pick the top line of the Drawing Title box.

 4. Pick a point above the line, as shown in the illustration. Press the Enter button on the mouse to end the Offset command.

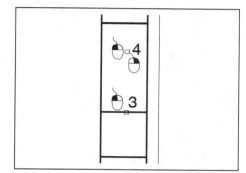

Make the First Revision Line

 1. Pick the OFFSET tool on the Modify tool box.

 2. Type <u>4'</u> and press the [Enter] key.

 3. Pick the top line of the Drawing Title box again.

 4. Pick a point above the line, and press the Enter button on the mouse to end the Offset command.

 5. Select the EXPLODE tool on the Modify tool box.

 6. Pick the new line, as shown in the illustration.

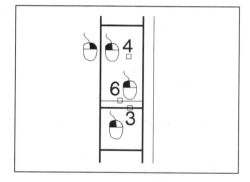

● ●

Changing a Line's Color Independent of the Layer

Exploding any Polyline creates a Line object instead of a Polyline with zero width. To make sure that it plots as a thin line, it may be necessary to set it to a color that is standard in your office for plotting light line weights. This is easy to do:

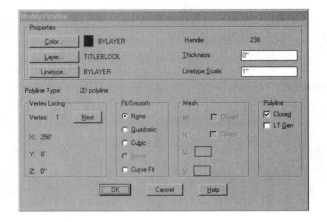

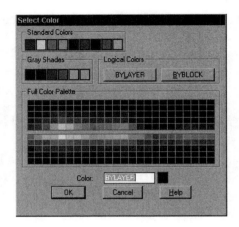

Pick the Properties (DDCHPROP) button on the tool bar. In the Modify Polyline dialog, pick the Color box.

Pick the color you want for the line in the Select Color dialog and pick OK.

Pick OK in the Modify Polyline Dialog, and the task is completed.

Create the Other Eleven Revision Lines

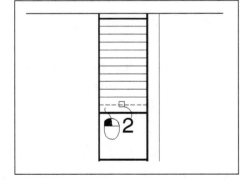

1. Pick the ARRAY tool from the Draw tool box.

2. Type R, followed by [Enter] to select the Rectangular Array option.

3. Pick the line we just created as shown in the illustration.

4. Type 11 and [Enter] for the number of rows to create.

5. Press the [Enter] key again to select the default of one column.

6. Type 4' and [Enter] to set the distance between the lines (4'-0" will be 1/4" when the drawing is plotted at a scale of 1/8" = 1'-0"). AutoCAD will quickly fill the box with ten copies of the line.

When to Use OFFSET, ARRAY, and COPY

All twelve lines should now be drawn as illustrated. We could have used OFFSET to create the lines, but in this case ARRAY shaves at least five seconds off the process. As a general rule, use OFFSET to duplicate four or fewer lines or objects, and ARRAY to make copies of five or more objects at regular intervals. COPY with the Multiple option selected is very efficient at duplicating objects that snap to irregularly spaced points accessible by an Osnap setting, such as Intersection. An example of this use of COPY is where the building structural grid is nonuniform; we can quickly COPY a single column symbol to all the grid intersections using the Multiple option of the COPY command.

Making the Logo Box

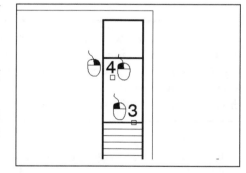

 1. Pick the OFFSET tool on the Modify tool box.)

 2. Type <u>40'</u> and press the [Enter] key.

 3. Pick the top line of the Drawing Title box again.

4. Pick a point above the line, and press the Enter button on the mouse to end the Offset command.

Inserting Your Company's Logo in the Title Block

Although Release 13 of AutoCAD allowed the insertion of bitmap images in a drawing, it did not allow easy scaling, duplication, and manipulation of the image's properties the way Release 14 does. In addition, Release 14 supports a greater number of bitmap file formats than did 13. In this next section we will see how to quickly insert a properly prepared company logo (mine, included on this book's companion disk.)

Properly prepared are indeed the operative words for bitmaps inserted into AutoCAD drawings. AutoCAD insists on enclosing all bitmaps in frames, which must be turned off if you don't want them to plot. This framing "feature" means that asymmetrical graphic designs such as the Partners

By Design logo suddenly get a look they never had on your letterhead. Depending on the software you use to create the logo, defining the area enclosed by the frame may be possible within your software, or you may have to let AutoCAD arbitrarily make the frame for you.

I created the Partners By Design logo in Corel Draw version 3.0a, and after several false starts, I was able to draw a frame around it that would coincide with AutoCAD's frame definition. I found that it was very important to export only the parts of the logo itself to the bitmap file. Drawing the frame allowed me to check the "Selected Entities Only" box in Corel Draw once the logo and frame were selected, thus setting the parameters of the image myself, rather than letting AutoCAD crop the image. Once the bitmap file is created, inserting it is pretty painless, as I shall explain below.

Plotter Driver Problems

Note: Most Windows 95/NT plotter drivers do not support combined raster and vector output (ours is brand-new, and it doesn't). Some have special drivers for plotting drawings created with programs such as Adobe Illustrator or Corel Draw, and you may be able to use these with special care when assigning scale factors. Until new Release 14 drivers are available, this will continue to be a problem.

ZOOM in on the Logo Box area so that it fills the screen. Drop down the Layer List from the tool bar and pick the Graphics layer to make it current.

Release 14's New Raster Image Tools

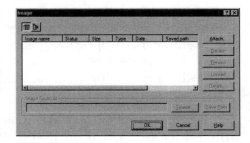

1. Pick the Insert menu on the tool bar and select Raster Image.

2. The Image dialog box will appear. Pick the <u>A</u>ttach button. Pick the Browse button and navigate to the floppy disk drive on your system.

3. Select the file PBDlogo.tif and pick the Open button.

4. Pick OK on the Image dialog box. You should now be back in the drawing, as shown in the illustration.

5. Pick a point to the left side of the upper third of the Logo box.

6. Type <u>200</u>, followed by [Enter] when prompted for the scale factor. Note that this is just the scale factor that works with our logo; your mileage may differ. The logo and its frame should now be inserted as shown in the illustration. If the logo is overlapping part of the title block, use the MOVE tool in the Modify tool box to position it as illustrated.

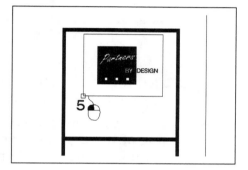

Stretching a Raster Image to Fit the Title Block

1. Pick the logo image by clicking on any part of it, thus turning on its Grips mode.

Reminder: When using Grips, you don't need to hold down the pick button to perform a drag operation as in some other Windows 95 and Windows 3.x programs. You just click on the Grip to highlight it, then click on it to "anchor" it to the mouse. After that, it will follow every move you make. Pressing the [F8] key will toggle Ortho mode on and off as you position the Grip point.

2. Pick the upper-left-corner Grip and drag it to align with the top line of the Logo box. Press the Pick button on the mouse to anchor it. It doesn't matter if you overlap the border line a little; just be sure to not leave the image frame visible inside the border line.

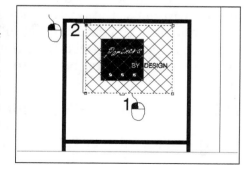

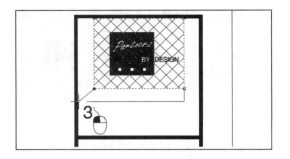 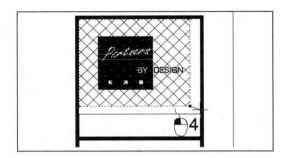

3. Pick the lower left Grip and drag it to the left border of the Logo box. Press the pick button on the mouse again.

4. Pick the lower-right-corner Grip and drag it to the right border line of the Logo box. Press the mouse's pick button to anchor the image inside the border line.

Making the Frame Invisible

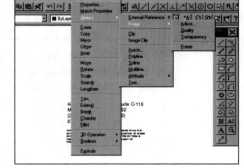

1. Pick the logo frame if the image is not already highlighted. Pick the Modify menu.

2. Pick Object, then Image, then Frame as shown in the illustration.

3. Type OFF followed by [Enter] to make the frame disappear.

At this point, before changing the Polyline's width, it would be a good idea to use the OFF-SET tool to make the box for our address and copyright notice. Pick the OFFSET, set the distance to about 14 feet, pick the polyline, and pick a point below it. Press the mouse's Enter button to end the command.

Using Edit Polyline to Change the Line's Width

1. Pick the Edit Polyline (PEDIT) tool on the Modify tool box.

2. Pick the polyline we just moved.

 3. Type <u>W</u> and press [Enter]. Type <u>4</u> and press [Enter] to change the line's width from 8 inches to 4 inches.

4. Press the [Enter] key again to end the Edit Polyline command.

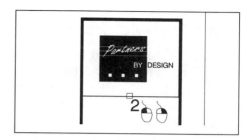

Add the Company Address and Copyright Notice

Well, we have a logo and an empty box. I simply picked the Paragraph Text tool (MTEXT) and created a window for my company's address and phone number (text height set at 1 foot), and then repeated the process to create the copyright notice at a text height of 4 inches. This illustration shows the final product.

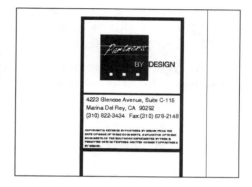

The final task is to save the drawing by picking the Save File button on the tool bar! Here's the completed title block sheet, ready for Saving As "Title_block dwg" or any other name you feel is appropriate.

But Where's the Title?

Oh, that! We could, of course, put in the project name and address if we wanted, and save the title block with a project-specific name such as "tutor-tb.dwg" (<u>tutor</u>ial building, <u>t</u>itle <u>b</u>lock). I prefer to put that information in the drawing itself as an added visual clue that confirms the drawing contents. It's really a matter of personal preference. The drawing number and name should always be part of the individual drawing file and so will never appear on the title block reference drawing. The same goes for issue dates, revision text, etc.

● ●

Other Possible Uses of Raster Images in Drawings

Release 14's ability to import raster images has other exciting possibilities. There are many times in remodel projects that I want to include photographs as part of a construction drawing, because often it's a lot easier to show a picture of a problem to be repaired, alongside an explanation of what has to be done, than it is to try to draft or verbally describe an existing condition. Other uses come to mind as well, especially the use of aerial photographs combined with site plans, and the use of photographs as demolition instructions ("Remove this stuff without disturbing these things!").

I encourage you to experiment with what AutoCAD calls **compound drawings**. The possibilities are intriguing.

DESIGN
DEVELOPMENT

PART TWO OVERVIEW

DESIGNING WITH AUTOCAD RELEASE 14

Contents

It's a sad fact, but in almost every firm that tries to use AutoCAD for design, the time required to get to the end of what should be a familiar process increases and the presumed efficiencies of developing design on CAD are never realized at the construction drawing phase. Instead, the time available for producing CDs is forced to be compressed, without a corresponding reduction in the work required.

• •

The Lure of the Computer Rendering

I recognize that every designer (including me) dreams of generating photorealistic renderings just by sketching shapes into the computer and having it magically supply the tedious detail and correct perspective in a real 3D model that we can then walk around and through. Our main desire is to get a design to look right, from inside and out, and to be able to visualize it with a sharp enough level of detail so that everyone involved can agree that the design is perfect (and wonderful).

Intellectually, we all believe that once we have the building *looking right*, it is a small step to engineering and construction drawings that will maintain this *rightness* (Wrightness?) without too many compromises. We all dream of "handing over" our 3D computer model to others who will swiftly develop final 2D drawings for construction.

Those who have developed 3D models in AutoCAD in order to produce renderings from them know the sad truth: When construction drawings are started, the building needs to be redrawn completely from a clean sheet in 2D. The model is too complex and unwieldy to use except for some of its geometry. By the time this process is complete, all profit has disappeared from the job because the firm has just paid to draw the entire building twice, in considerable detail.

The pursuit of the computer-rendered design has made many designers forget that CAD stands for Computer Aided Design/Drafting. Instead of using CAD software to draw design at the conceptual level, we will explore how to use its considerable power as a geometry engine to aid in conceptual design and final design development while avoiding the pitfall of spending 50 percent of the project time producing a rendering.

When and How to Use CAD for Design

Our first step will be to use AutoCAD to establish the site boundaries by drawing property lines in 3D. We will then use AutoCAD's 3D drawing tools to create a perspective and isometric grid for us to use in hand-drawing design concept sketches.

Load the a-1_blank.dwg drawing you created in Chapters 3 and 4, or load it from the sample files included on the companion disk.

Drawing the Perspective and Isometric Grid for Manual Sketching

Drawing the First Orthogonal Property Lines

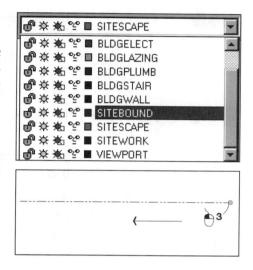

ORTHO 1. Turn Ortho mode on. Select the Sitebound layer from the drop-down Layer List on the tool bar at the top of the screen.

2. Select the Line tool.

3. Pick a starting point in the upper right corner of the screen, leaving enough room for dimensions and notes.

4. Drag the line 180 degree to the left and type <u>235'</u>, then press the [Enter] key or the Enter button on the mouse.

This line is on the highest elevation of the property. The lower edge (in plan and in contour) of the property is 3'-0" below this elevation.

Direct Distance Entry Explained

Typing the distances and pressing [Enter] or the mouse's Enter button is called Direct Distance Entry. It is a distance entry method far superior to AutoCAD 12's requirement of typing the "@123'<180" vector specification. Direct Distance Entry has been available on AutoCAD LT since version 2.0, but it must be rembered that it is best used with Ortho mode on.

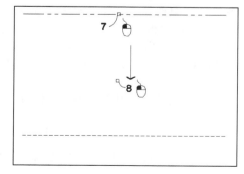

5. Pick the OFFSET button.

6. Type <u>145'</u>, and press the [Enter] key when prompted for the offset distance.

7. Pick the property line at the top of the screen.

8. Pick any place below the upper property line. The new lower prop-

erty boundary will appear at the bottom of the screen as indicated by the dashed line in the illustration.

 9. Type <u>CHANGE</u> and press [Enter]. AutoCAD will prompt you to "Select Objects."

 10. Pick the lower line and press the Enter button on the mouse, or press [Enter] on the keyboard.

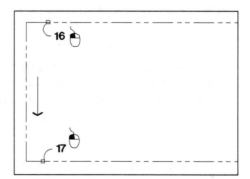

 11. Press the [P] key and [Enter] to change properties.

 12. Press the [E] key and [Enter] to change the elevation of the line.

13. Type <u>-3'</u> and [Enter] to move the line 3'-0" below the line at the top of the screen.

14. Press the [Enter] key again to end the command. Keep Ortho mode on.

Draw the Three-Dimensional Vertical Property Line

 15. Pick the LINE tool.

 16. Pick the ENDPOINT of the top Property line.

 17. Pick the ENDPOINT of the lower line and press the Enter button on the mouse.

This step creates a line with its finishing endpoint 3'-0" lower than its starting endpoint. The distances in our example are shown on the design sketch and are typically taken from surveyors' data or official government maps.

Draw the Angled Property Line

 1. Select the OFFSET tool.

 2. Type <u>210'</u> and press [Enter].

3. Pick the left vertical line you just drew.

4. Pick a point to the right of the line. Press the Enter button on the mouse to end the command.

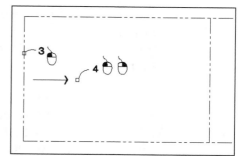

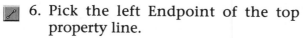

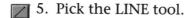
5. Pick the LINE tool.

6. Pick the left Endpoint of the top property line.

7. Drag the line down and pick the Endpoint of the vertical line you just Offset. Press the Enter button on the mouse to end the Line command.

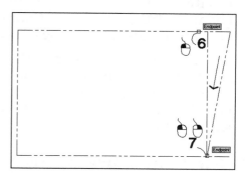

Note: The new angled property line is lower by 3'-0" at its intersection with the bottom horizontal line than at the top. If we had used the Intersection Osnap to pick the last point for the line, AutoCAD might have returned a result that would have appeared to create an intersection properly but in reality could have been several inches away from the desired point. This often happens when Ortho mode is active.

Trim the Lower Three-Dimensional Property Boundary

1. Pick the TRIM tool.

2. Pick the diagonal property line and press the Enter button on the mouse.

3. Pick the overlapping end of the lower property line. This trims the property line to end precisely at the diagonal right boundary. Press the Enter button on the mouse to end the Trim command.

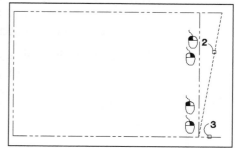

4. Press the [E] key or pick the ERASE tool. Pick the vertical construction line and press the Enter button on the mouse.

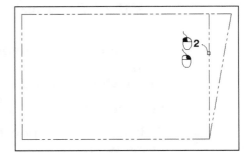

We have completed the 3D boundary for the site. The method we have used, employing the old CHANGE command, is faster by several minutes than drawing in AutoCAD's 3D coordinate system, which requires constant changes of the UCS (User Coordinate System), and calculation of specific coordinates or drawing of several lines that must later be erased.

We are now going to overlay the site with a grid drawn at the same level as the upper boundary line elevation (0'-0"). The grid will be 10' × 10', constructed using the ARRAY command. We will use the grid for reference in hand-drawn design studies, overlaid on an isometric plot of the site and grid.

Draw the 10' × 10' Foot Grid

 1. Select the Sitescape Layer from the pull-down layer list, making it the current layer. Make sure Ortho Mode is on (press [F8] or double-click the Ortho button).

 2. Pick the LINE tool.

 3. Pick the upper-right **Intersection** of the property boundary lines. Drag the line to the left, past the left vertical property line, and press the Enter button on the mouse to end the Line command.

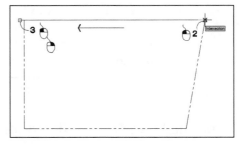

 4. Pick the LINE tool again or simply press the Enter button on the mouse to recall the last command, and pick the upper right Intersection again.

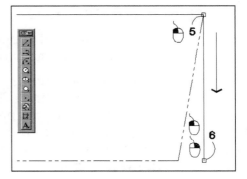

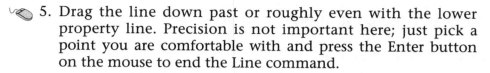

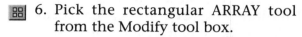 5. Drag the line down past or roughly even with the lower property line. Precision is not important here; just pick a point you are comfortable with and press the Enter button on the mouse to end the Line command.

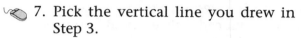 6. Pick the rectangular ARRAY tool from the Modify tool box.

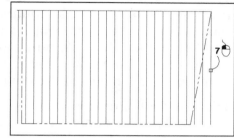

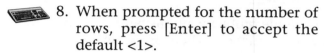 7. Pick the vertical line you drew in Step 3.

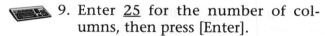 8. When prompted for the number of rows, press [Enter] to accept the default <1>.

9. Enter 25 for the number of columns, then press [Enter].

10. Enter -10' for the prompted distance between columns; press [Enter]. This will create 24 copies of the line 10' on center from right to left. Entering a positive 10' would draw the lines to the right of the starting position.

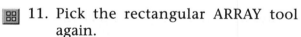 11. Pick the rectangular ARRAY tool again.

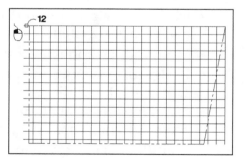

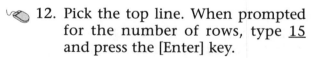 12. Pick the top line. When prompted for the number of rows, type 15 and press the [Enter] key.

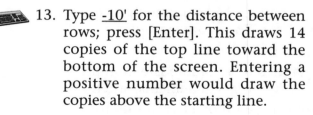 13. Type -10' for the distance between rows; press [Enter]. This draws 14 copies of the top line toward the bottom of the screen. Entering a positive number would draw the copies above the starting line.

Our design grid is now complete. We are a couple of feet short of the lower property line, but that's the lower end of the parking lot, so it doesn't matter too much.

• •
Designing the Tutorial Building

The Tutorial Building's Purpose

The goal of the Tutorial Building design is to create a structure that will incorporate all the tasks required to draw a set of construction documents, not to create an award-winning architectural design. *That's your job*. Remember: To *build* an award-winning design, you must produce construction drawings. They must be produced efficiently at profit, so you can afford to design the next award winner.

The Tutorial Building is intended to include problems not usually encountered in Type 5 (residential housing) construction. There are several residential architecture add-on programs that help in producing Type 5 construction drawings. The nature of commercial building drawing is very diverse, ranging from multibuilding complexes on hundreds of acres to more mundane industrial park structures or single-story retail buildings.

The Tutorial Building Program

Out of this diversity, we selected elements that occur in most types of commercial structures:

- Floor plans that are similar from one to the next, but not exact replicas of the floor below
- Exit stairs, elevators, and multifloor penetrations of mechanical and electrical systems
- Multistall rest room layouts
- Design of systems and spaces that comply with the ADA (Americans with Disabilities Act), within the building and on the site
- Flat roof systems that must support HVAC equipment, incorporate access hatches, roof drains, and overflow scuppers, etc.
- Multifloor glazing systems (curtain walls)
- Interior spaces that must be planned to accommodate multiple uses, with a variety of ceiling systems, lighting requirements, and finishes
- Structural systems not commonly employed in residential construction

The program we developed was based on a low-rise commercial structure, because I was fairly confident that no one would want to draw forty floors of a high-rise just for the experience. I decided on two floors: steel frame with metal deck and lightweight concrete floor for the upper level, metal deck and built-up roof, and concrete-slab lower floor, on a poured-in-place concrete foundation. The building is theoretically located in a suburban "business park" zoned for commercial and light manufacturing use.

The building gross leasable area will be about 17,000 square feet on the first (ground) floor, and 16,000 square feet on the second floor. A thousand square feet is allocated for a common-area lobby on the first floor. The standard office use occupant load of 100 square feet per person produces a maximum building capacity of 220 people, making it ideal to lease to either a single-service business tenant or several smaller tenants. I included an on-site parking area for the purpose of dealing with parking layout drafting problems, not because of any Planning and Zoning mandate!

The Imaginary Building Site

Imagined setback requirements are 10 feet, which is common in business parks where 20-foot separation between structures allows adequate light; they are coupled with parking requirements designed to create a "campus-like" setting. A quick study showed that the most efficient site layout would concentrate parking on one long side of the site and use the 10-foot setback to define the building perimeter.

Developing Concept Design Sketches

The first sketches were simple building shape studies to see what kind of mass would work best on the site, while still meeting the program. The site has a problem of being very narrow at the public street side (on the east), where recognition of the building is most important (to my imaginary client).

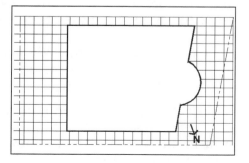

Siting and Massing Studies

Here's one of the first ideas, with the rotunda a prominent feature of the street side of the building. I actually did this on a tissue overlay of the plotted grid and re-created it on Auto-

CAD to see if it could be done quickly by using a Polyline and stretching it into the desired shape(s). It worked, but tissue and pencil are still faster, at least for me.

This design has interior problems: Its square shape requires two main corridors wrapping around a central core to create leasable perimeter spaces. I will not have enough core elements (rest rooms, elevators, etc.) on the first floor to make this work.

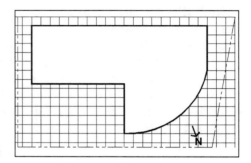

The second iteration has a long street-side front, and parking on the northwest corner. It works well enough, but the core service elements need to be in the widest part of the interior, which is unfortunately toward the front. That makes for a long hike to the elevator and rest room by a tenant at the back of the building.

The final solution was, like almost everything in life, a compromise. The long rectangle allows a single interior corridor and central core. We manage to get the entrance rotunda on the northwest corner adjacent to both the street and parking, and yet cover half the street side of the site with building facade.

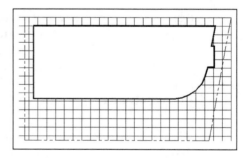

Yes, I got straight A's in all my rationalization classes in college.

Isometric Design Studies

In the initial design sketches, I turned the grid to give more of a street-front view of the building while allowing myself to see the long side.

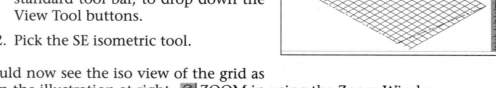

 1. Pick the Names Views button on the standard tool bar, to drop down the View Tool buttons.

 2. Pick the SE isometric tool.

You should now see the iso view of the grid as shown in the illustration at right. ZOOM in using the Zoom Window tool on the tool bar. Once we have the view to our liking, we can plot it and use it as the basis for isometric design studies.

The first-pass sketch is a rough study of the basic geometry of the building. The site grid allows for easy spatial reference when sketching.

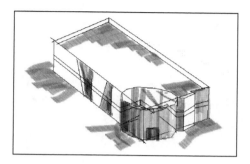

The next sketch shows a lobby "rotunda" with a skylight and a more developed idea of the fenestration on the other surfaces.

The final study omits the skylight for cost reasons and covers the rotunda with a metal roof over clerestory windows located along its two straight sides.

I really debated with myself on the entrance rotunda idea but in the end caved in and included one in the building. You should understand that I was going to do an entire chapter on the fastest way to draw all the current design clichés, but there are so many of them that it would have consumed most of the book. In the end, I threw this one in because it is so popular in this building type here in southern California. Oh, well . . .

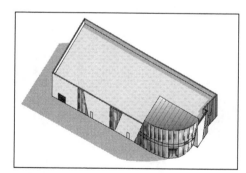

From these illustrations, I hope you see how design can become Computer Aided Design, and how the computer can be a valuable aid in getting the process started, just by doing a site layout. With the completion of these manual sketches, we are ready to proceed to more defined design development: preliminary plan drawings, design elevations, and schematic sections.

CHAPTER SIX

BUILDING WALL AND CURTAIN WALL LAYOUT, FIRST FLOOR

Contents

In this chapter we will be using the property lines to locate the building relative to the overall site boundaries, creating a Multiline style for our building wall, and other tricks of establishing the preliminary building geometry on the site and drawing the preliminary curved curtain wall layout.

Basic Technique

Having drawn the property boundary lines, we will use the OFFSET command to establish required setback on the west (angled) end of the site. From this point we will use Direct Distance Entry and Running Object Snap Intersection to draw the walls using Release 14 Multiple Lines (MLINEs).

This method will allow us to quickly "sketch" the building plan for the preliminary planning to follow. All the following methods work equally well for final construction drawing of plans as well.

Note: The Tutorial Building's site is small enough to fit on a 24" x 36" sheet when plotted at 1/8" to 1'-0" scale. This makes it easy to use one drawing file for setting up both site and building plan drawings in the Design Development Phase.

Using Zoning Setbacks to Locate Building Walls

Required setbacks for this site are a minimum of 10 feet. We are going to also offset the north boundary the distance required to create the visitor parking area. Follow the steps illustrated below to create construction lines that will give us Intersections to pick with Running Object Snap (Osnap) when we are drawing the building shell walls.

Note: If you are using AutoCAD LT 95, you will need to change the elevation of the lower property boundary line to 0'-0" using the CHANGE command as explained in Chapter 5. You will also need to use the ▥ PROPERTIES button to change the Z coordinate for each of the ends of the right and left property lines to 0'-0". To do this, pick the Properties button on the tool bar, and pick one of the lines. Find the end or start point in the Z coordinate window of the dialog box that is at -3'-0" and enter the value of 0 for it. Pick OK. Repeat for the other line. AutoCAD Release 14 users can set the Apparent Intersection Osnap instead of making the elevation changes. Apparent Intersection is capable of handling overlapping lines that are on different 3D elevations.

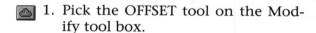

1. Pick the OFFSET tool on the Modify tool box.

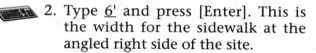

2. Type <u>6'</u> and press [Enter]. This is the width for the sidewalk at the angled right side of the site.

3. Pick the right side property boundary. Pick a point to the left of the line. Press the Enter button on the mouse to end the Offset command.

4. Press the Enter button on the mouse to recall the Offset command, or pick the OFFSET tool again.

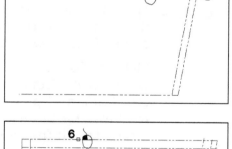

5. Type <u>10'</u> and press [Enter].

6. Pick the top boundary line. Pick a point below it.

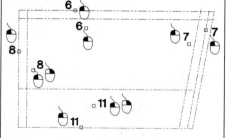

7. Pick the right (angled) sidewalk boundary line. Pick a point to the left.

8. Pick the left property boundary, then pick a point to its right. Press the Enter button on the mouse to end the Offset command.

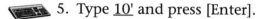
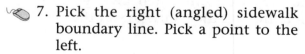
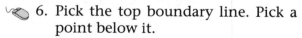
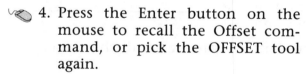
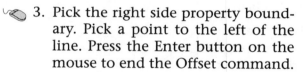

9. Press the Enter button on the mouse to recall the Offset command.

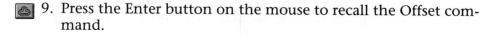

10. Type <u>48'</u> and press [Enter]. This is the distance from the lower property boundary to the lower building wall, as well as the overall dimension of the parking lot.

11. Pick the lower property boundary. Pick a point above it. Press the Enter button on the mouse to end the Offset command.

Creating the Remaining Construction Lines

The next step is to OFFSET construction lines for the curved glass wall of the entrance lobby and for the upper horizontal wall of the rectangular projection, which the top of the curved glazed wall dies into. These will be the last of the construction lines we need.

 1. Pick the OFFSET tool.

 2. Type <u>38'</u> and press [Enter]. This is the radius of the ground level curve of the lobby rotunda glazing.

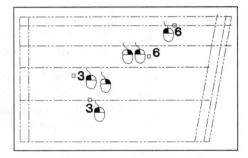

3. Pick the lower wall line (the one we offset <u>48'</u> from the lower property line), and pick a point above it. Press the Enter button on the mouse to exit the Offset command.

4. Press the mouse's Enter button again to recall the Offset command.

5. Type <u>24'</u> and press [Enter]. This is the line from the top left corner of the building to the start of the rectangular projection on the angled side.

6. Pick the upper wall construction line, then pick a point below it. Press the Enter button on the mouse to exit the Offset command.

ORTHO 7. Turn Ortho Mode on by double-clicking the button at the bottom of the screen or by pressing the [F8] key.

8. Pick the LINE tool from the Draw tool box.

9. Pick the Intersection (or Apparent Intersection) of the angled wall construction line as shown. Drag the line down past the

lower building construction line, and pick again to anchor its end-point. Press the Enter button on the mouse to end the Line command.

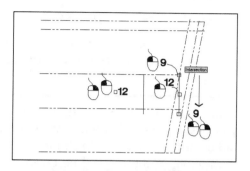

10. Pick the OFFSET tool.

11. Type <u>38'</u> and press [Enter].

12. Pick the vertical line we just drew, and pick a point to the left of it. Press the Enter button on the mouse to end the Offset command. You can now use the ERASE tool on the Modify tool box to get rid of the original line.

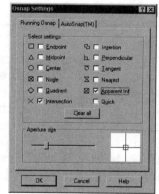

We are now ready to draw the building walls for the first floor. If you have not done so, set Intersection *and* Apparent Intersection as your Running Osnap.

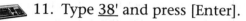

Drawing the Outside Radiused Glazed Wall

1. Select the Bldglazing layer from the drop-down layer list, making it the current layer. Turn the Sitework layer off to make the grid disappear. Make sure Ortho mode is on.

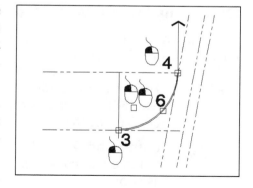

2. Select the ARC/Start End Angle tool from the Draw tool box, or from the Draw menu if you have not placed this tool in your Draw tool box.

3. Pick the lower left Intersection, as shown.

4. Pick the upper right Intersection. Drag the cursor up at 90 degrees until the arc is formed properly, then press the pick button on the mouse.

5. Pick the OFFSET tool. Type <u>8</u> and press [Enter].

 6. Pick the arc you just drew, then a point inside the building area to create the inside glazing line.

Note: Drawing arcs with Ortho on is the best way to ensure that they have the radius you desire. This is important when drawing a partial circle of known (or required) radius. It is important to pick the arc Start and End points in the order shown in the illustration, or the arc will be drawn in the opposite direction—up and to the left—instead of down and to the right.

To design an irregular curved wall or curtain (glazed) wall, two methods may be employed. The more rational is to draw a series of arcs using the ARC 3 Point tool. The other method is to draw a number of straight Polyline (Pline) segments, then convert them to curves using the Spline option in the Polyline Edit tool/command. The downside to using Polylines is that their curves cannot be dimensioned for radii. You will need to deal with the issue of drawing something that cannot be easily communicated to curtain wall manufacturers and contractors except by exchange of CAD files and full-size plotted templates. Here's how to use both methods.

• •
Designing Curved Walls Using Arc Segments

Drop down the Layer list from the Object Properties tool bar and turn Bldglazing off; then pick a new Current Layer such as Bldgwalls.

`ORTHO` 1. Turn Ortho mode off by pressing the [F8] key or double-clicking the ORTHO button.

2. Using the ZOOM Window tool, enlarge the curved glazed wall area as shown.

3. Pick the ARC 3 Point tool from the Draw tool box.

4. Pick the lower left Intersection of the curved glazing area.

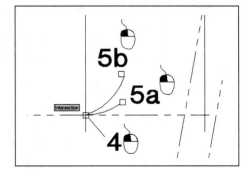

5. Pick a point about 10 feet away (shown as 5a in the illustration). Drag the arc until it covers about a third of the space between the ends of the curved glazing, and press the pick button on the mouse at point 5b.

The exact geometry of the arc doesn't matter at this point. Think of it as drawing freehand curves. We'll draw two more arcs:

6. Press the Enter button on the mouse to recall the ARC 3 Point command.

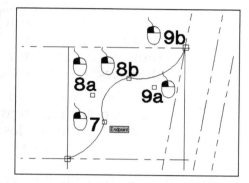

7. Pick the **Endpoint** of the free end of the first arc.

8. Pick a point about 10 feet away (point 8a), and drag the new arc into position. Press the pick button on the mouse at point 8b to anchor it.

9. Repeat these steps to draw the third arc, starting at the **Endpoint** 8b, but anchor its end at the Intersection of the two construction lines at point 9b, as shown.

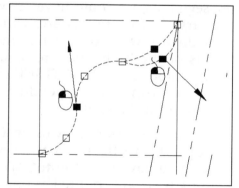

Now for the fun part of this freehand design: Turn on the Grips for all three arcs by picking each one in turn, or by using a Crossing box.

10. Pick the Grips at the intersections of the arcs and drag them into the desired positions. Use the Grips at the arcs' midpoints to adjust the arcs' radii. When you are satisfied with the design, press [Cancel] twice to turn the Grips off.

11. Pick the OFFSET tool. Type <u>8</u> and press the mouse's Enter button.

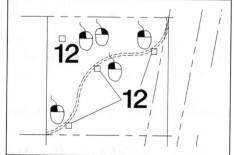

12. Pick the arcs, and pick a point in the building interior for the offset

direction. Press the mouse's Enter button to end the Offset command.

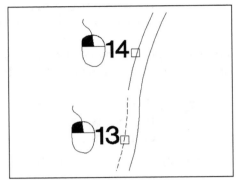

13. Zoom in closer and look at the ends of the new (offset) arcs. You will notice that they are slightly separated at the endpoints. Arcs that are OFFSET in any direction must be rejoined using the FILLET tool/command, with the Fillet Radius set to zero: Pick the FILLET tool on the Modify tool box and then pick one of the arcs.

14. Pick the other arc on its adjacent end. Repeat these steps for the other arcs. You don't need to be ZOOMed in so close to make Filleting work; it is important in this example for illustration purposes.

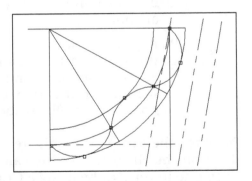

To make a more controlled design of regular arc segments, we could have drawn them over our first 38-foot radius arc, which we could have divided with construction lines. The midpoints of the individual arcs could also have been snapped to the OFFSET copies of the large arc, using the Perpendicular Osnap, as shown in this illustration.

This design was started by using a Polar ARRAY to copy the vertical construction line three times (you have to tell AutoCAD to make four copies because the original line is always counted as one of the objects in the ARRAY). Then we just OFFSET the original arc a distance of 4'-0" to create the rest of our construction lines, and drew the arcs for the curtain wall.

Unlike curved Polylines, arcs have true radii that can easily be dimensioned. Arcs are also easy to divide into glass segments when doing final construction drawings, as we shall demonstrate in later chapters.

Considerations in Working with Consultants

If you are sharing files with consultants (engineers, interior designers, landscape architects, etc.) who are still using AutoCAD Release 12, splined polylines will not transfer reliably to Release 12 format. Splined polylines in Release 13 and 14 drawings are Non Rational B-Splines (NURBS), which are only approximated, not supported, in Release 12. When you save an R 14 drawing as an R 12 drawing, the splined curve is transformed to an exploded polyline with a large number of individual segments.

If you accept a drawing from someone using Release 12, you will not be receiving your splines back in their original form. There may be subtle differences between the curves in the R 14 and the R 12 drawings, which will only be obvious when they are overlaid, combined, or referenced. With these caveats in mind, let's explore spline curves.

Drawing Curved Walls and Glazing Using Polyline Splines

First, we want to erase the drawing we have just done:

1. Pick the Erase tool from the Modify tool box, or press the [E] key and [Enter], then pick the arcs. Press the Enter button on the mouse when you are finished.

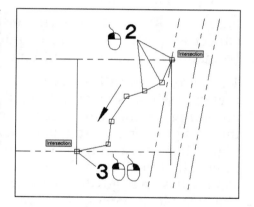

2. Pick the POLYLINE tool from the Draw tool box.

3. Pick the upper right intersection of the glazing area and drag the Poly-line a short distance, pick an end-point, and continue in this way to create a series of segments in the rough shape of the curve you want.

4. Pick the lower left intersection of the glazing area to complete the seg-ments. Press the Enter button on the mouse to end the Polyline com-mand.

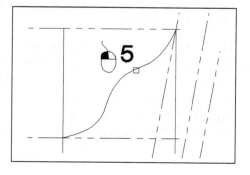

5. Pick the Edit Polyline tool (PEDIT) from the Modify tool box. Pick the polyline you just drew.

6. On the command line, AutoCAD presents us with a laundry list of editing options after the command prompt. We are only interested in Spline, so type <u>S</u> and press [Enter]. Press [Enter] again to exit the Polyline Editor. You will now observe a graceful curve in place of the original Polyline. If you click on it to turn its Grips on, you will notice that they are located at the positions of the original Polyline's joints. Practice dragging them around to change the Polyline's shape.

Drawing the Building Walls

Make Bldgwall the current layer by selecting it from the drop-down layer list on the Object Properties tool bar.

Setting Up a Multiline Style

Pick the  Multiline Style button in the Object Properties tool bar, or pull down the Format menu and select Multiline Style.

The cryptic Multiline Style dialog box will present itself, as shown in the illustration. We are going to add a new style, so swipe the mouse's pointer over the name STANDARD to highlight it. We can now type our style name: <u>bldgwall</u>. Multiline style names follow the same rules that govern naming blocks, and just like block names, they are automatically converted to uppercase (don't ask; it's a feature). Now press the *Add* button, and then the *Element Properties* button.

The Element Properties dialog now appears, and we can set up our simple 8-inch-thick building wall. In the properties window, there are two lines of data. The top one should be highlighted, but if it is not, pick it with the mouse. Now, in the Offset window, highlight

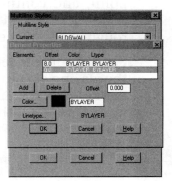

the value shown there and type <u>8</u> to replace it. Pick the next line of data in the properties list window and type <u>0</u> in the Offset window as its value. The dialog box should now look like the illustration on the previous page. Pick the OK button.

Now press the **Multiline Properties** button on the Multiline Styles dialog, and you'll see the Multiline Properties dialog. The first thing we want to do is to tell AutoCAD to automatically cap the ends of our wall line, so we check the **Start** and **End** boxes opposite the word **Line** in the Caps section.

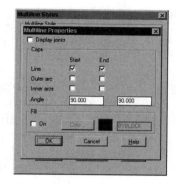

Notice the Fill section below Caps: This allows us to draw a filled multiline with any color fill and do it independently of the layer being used for drawing. This is an excellent way to poché (fill in) walls that are regular enough to be drawn straight or curved for long distances. The poché Multiline can be on a dedicated layer, which can be turned on or off as needed.

That's all we need to do here. Pick the OK button. Pick the OK button on the Multiline Style dialog.

The STANDARD Multiline style cannot be renamed, edited, or deleted. If you wish to modify a Multiline style you created and then drew with, you must rename it to change its properties, because a Multiline style used in a drawing cannot be altered.

What we have just done is designed an 8-inch-wide Multiline. When Multilines are drawn from right to left, they are offset in the 90 degree direction (up), and when drawn from left to right, are offset in the 270 degree direction (down). As we draw walls around the building in a clockwise direction, the offset wall line will always be to the **inside of the building**.

There are lots of things you can do with Multilines that I don't have space to cover here, but one obvious thing you may want to experiment with is creating an MLINE that automatically draws a window's glass line in the correct color while drawing the building walls parallel to it. Creating multilines gives us a whole new way to waste time.

Making the Preliminary Building Shell

We need to make one more construction line to locate the intersection of the 90 degree wall on the west end of the building. Once that's done, we can draw walls. **Make sure that your Running Osnap is set to Intersection.**

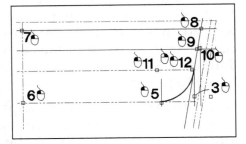

 1. Pick the OFFSET tool on the Modify tool box.

 2. Type <u>7'-4"</u> and press the Enter button on the mouse. Because the designer is insisting that the horizontal wall that the curtain wall dies into be 8'-0" long and end precisely at the junction with the curtain wall's inside arc, we need to subtract the 8" thickness of the curtain wall from our Offset distance—hence the 7'-4".

 3. Pick the original construction line, then a point to the right. Now for the wall.

 Note: We have used the Match Properties tool on the tool bar to change the construction lines to the Bldgwall layer and line type in order to pick intersections more easily.

---

 4. Pick the Multiline (MLINE) tool on the Draw tool box.

 5. Pick the Intersection of the lower outside curtain wall arc and the construction lines.

 6. Drag the cursor to the left and pick the Intersection of the two construction lines.

 7. Pick the Intersection of the upper wall and left wall construction lines.

 8. Pick the Intersection of the angled and top horizontal construction lines.

9. Pick the Intersection of the angled construction line and the next horizontal one.

10. Pick the Intersection of the horizontal and vertical construction lines as illustrated.

11. Pick the **Perpendicular** Osnap button on the tool bar, then pick the next horizontal construction line down.

12. Finally, ZOOM in and pick the Intersection of the **inside** arc of the curtain wall.

Multilines work fine with Direct Distance Entry, if you know what your dimensional targets are. Practice more with Multilines to draw interior walls. They also work well with Tracking to create door openings easily.

Now that we have the walls sketched out, we will use them to do some final perspective views of the building design, which we can plot and use as the basis for hand-drawn artistic renderings. This is our first 3D checkpoint of the design.

VISUALIZING THE DESIGN IN 3D

Contents

Using AutoCAD as a 3D modeling tool is very time consuming and requires learning the User Coordinate System (UCS in ACADspeak) interface at an expert level. Using AutoCAD's Dynamic 3D viewing power to set up design rendering underlays, on the other hand, is **easy** to do and requires little learning time. **In fact, you don't even need to know how to draft with AutoCAD to use it as a 3D visualization/design tool—** all you need to know is the material in this chapter (as long as someone does the drafting for you).

This chapter explores quick and easy methods to set up perspective views of the basic building shape that can be traced over manually or imported into computer illustration programs for digital rendering (not modeling!). The main idea of this chapter is to use the best tool for the job, with the goal being the best presentation art required to sell the project's design.

Good presentation art reflects the attitude and skills of your firm. It should communicate the type of vision that sets your firm apart from the rest of the competition. Look-alike computer renderings cannot do this. One ray trace looks just like another, regardless of the differences in design. Computer renderings of different designs could all have originated from one architect, as far as the layman's eye is concerned.

For this reason, and because 3D modeling with a CAD program is extremely tedious and hard to master, we recommend letting the computer do what it does best, and letting human artists do what they do best, **assisted** by the computer, whenever possible. Having said that, there are occasions where 3D modeling is essential in architecture, urban planning, and engineering. The expansion of Atlanta's airport several years ago is a good example: The 3D dynamic model of the flight paths and their relationship to surrounding neighborhoods sold the project to the community. However, for rendering a single building design, the manual process is still more efficient.

We will use our preliminary building outline to generate several 3D perspective views of the Tutorial Building, which we will use to see if the plan shape accomplishes our initial design goals. If it needs modification, now is the time to manipulate it before any more detail is added to the CAD file. Changes to more detailed drawings require more time in manipulating a greater number of CAD objects. Time spent this way is wasted. Time wasted equals profits lost.

Extruding the Walls into 3D Shapes

The first step in generating a 3D view of our building wall plan is to make the plan itself have a third dimension. We do this by extruding the walls in the Z coordinate (perpendicular to the plan). The Z coordinate dimension of an object is called its "thickness" in ACADspeak. Get used to the term. It's been a "feature" of AutoCAD since the introduction of 3D coordinates.

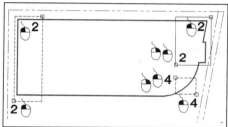

1. Pick the **Properties** button on the tool bar.

2. Using a crossing box, select all the walls except the curved one, as shown here. When you have highlighted all of them, press the Enter button on the mouse.

 3. You will now see the Change Properties dialog box, shown here. Pick the text entry box opposite <u>T</u>hickness: and type <u>35'</u>. Press [Enter]. Pick the OK button to return to the plan.

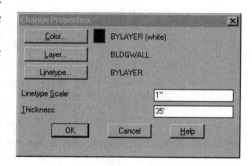

We designed the Tutorial Building to have a 16-foot floor to floor and floor to roof (slab to slab) separation, which takes us to 32 feet, plus an allowance for a 3-foot parapet, giving us 35-foot-high (thick!) walls.

4. Again using a crossing box, select all of the curved wall and press the Enter button on the mouse.

 5. In the Change Properties dialog box, type <u>39'</u>. Press [Enter].

This dimension will give us a 4-foot projection above the parapet for the clerestory at the curved wall. Of course, nothing appears to have changed on our drawing. To see the changes we will use AutoCAD's 3D viewing tools.

Creating Perspective Views of the Building

1. Pick <u>V</u>iew on the menu bar at the top of the drawing screen, then pick 3D D<u>y</u>namic View.

2. At the Select objects: prompt, select the entire building with a Window. It's not necessary to select the property boundary as well.

3. At the "CAmera/TArget/Distance/ POints/PAn/Zoom/TWist/Hide/ Off/Undo/<eXit>:" prompt, type <u>PO</u> and press [Enter]. We are now going to point out to AutoCAD what we want to look at and where we want to stand relative to what we are viewing. Pick the ✎**Nearest** Osnap button and then pick a point near the middle of the curved glass wall.

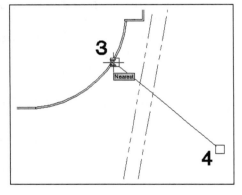

4. Pick a point to the lower left of the curved glass wall and press the Enter button on the mouse.

5. You will now find yourself looking at a parallel projection of the building elevation. The next step is to turn on perspective viewing.

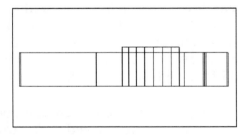

6. Type <u>D</u> and [Enter] to turn on perspective viewing and to activate the Distance function. Move the mouse to slide the building farther away and closer. Notice that the slider bar at the top of the screen moves to give you a graphic indication of your relative position. When you are satisfied with the distance setting, press the pick button.

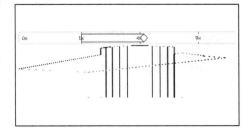

7. Type <u>H</u> and [Enter] to see the building with the hidden lines removed, as shown here.

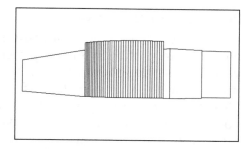

8. Type <u>CA</u> to activate the CAmera. Move the mouse slowly up and down to adjust the eye level until it is above the ground plane. If the building spins out of orientation, slowly move the mouse until it returns to the proper view. Type <u>PA</u> to activate PAn, and move the building down on the screen until it is more centered. Type <u>H</u> to hide the lines in the background.

You should now be looking at a view similar to the one in the illustration. To adjust it further, you can use the Zoom command (type <u>Z</u> and [Enter]). The Zoom command in Dynamic 3D View does not work like AutoCAD's normal zoom command. It should actually be named LENS because it allows you to dynamically change the lens you are looking through, as if our virtual camera had infinitely variable and interchangeable lenses.

Saving the View

Pick the <u>V</u>iew menu from the top screen menu. Pick <u>N</u>amed Views. Pick the **New** button on the View Control dialog box. You will see a second dialog box, as shown in the illustration. Type in the name of the view: Use a name that you will easily recognize, and don't skimp on words. A good name might be "perspective_NW_corner". Do not use spaces or periods in the name.

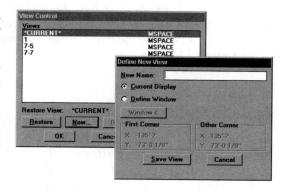

When you have typed the new name, save the view by picking the **<u>S</u>ave** View button, then picking the **OK** button on the View Control dialog box. Now save the drawing, as a precaution.

Plotting the Perspective

To use the perspective as an underlay for hand-drawn renderings, you will need to plot the view on paper. As we have pointed out before, how you do this depends on your system setup. If you are an individual practice with your own large format plotter, or if you use an outside plotting service, or if you work for a large firm with a networked plotter, your plot setup and use of third-party software will dictate the steps you must follow. The simplest way to plot is to the Windows 95 or NT system plotter/printer. When doing this, I almost always plot from a Window that I select depending on how much space I want around the perspective image.

Plot scale should be chosen first to fit on the size paper you will use for the actual rendering. It is usually easier to "play" with this variable by selecting different paper sizes and plot scales than by using AutoCAD's "plot to fit" check box. Always try a full preview before plotting. I don't spend any time twiddling with line weights. This plot is not a work of art, just the foundation for the final artwork.

Don't just settle for one view. Go ahead and experiment with different views. You can select a new series of POints at any time as well, and create some dramatic effects with different Zoom factors. To return to the parallel projection (turn perspective viewing off), type X for eXit. To return to the Plan View, pull down the View menu and select 3D Viewpoint > Plan View > World UCS.

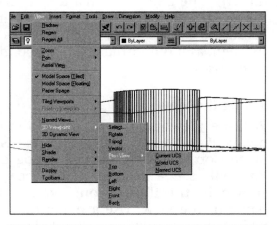

● ● ● ● ● ● ● ● ● ● ● ● ●

Evaluating Design

The illustration on the right shows the maximum slant we will get at the top right side of the curved glass wall. This is not as radical a slope as we had hoped for. To increase the slope, we will need to move the curved glass wall farther back from the street side of the building. Once that is done, we can proceed with adding

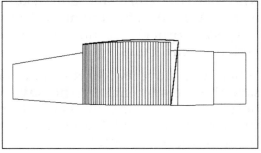

the structural columns, entrance doors, and exits for the first floor. Then it will be time to do design elevations and sections.

Changing the Plan at the Entrance Rotunda

Return to the plan view using the command sequences noted earlier in this chapter.

1. Make sure Ortho Mode is on. Pick the MOVE tool from the Modify tool box.

2. Select the curved glass wall and the ends of the horizontal building walls it connects to with a Window. First pick a point in the upper left positioned so that the Window will cover all three entities without covering the short horizontal wall at the top.

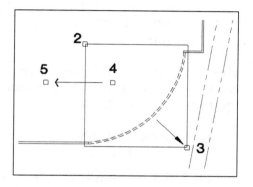

3. Drag the cursor to the right and down. Press the Enter button on the mouse.

4. Pick a point to the left of the curved wall. With Ortho mode on, drag the wall to the left.

5. Type <u>4</u>' and [Enter] or press the Enter button on the mouse.

The wall will now be 4 feet to the left of its previous position.

Note: This method of Direct Distance Entry also works with the COPY command.

Revising the Building Walls

1. Pick the TRIM tool from the Modify tool bar.

2. Pick the upper left corner of a Window to select the end of the horizontal wall, as shown in the illustration.

3. Pick the lower corner of the Window.

4. Type <u>F</u> for the FENCE selection method. Press the [Enter] key or the Enter mouse button.

5. Pick a point below the ends of the horizontal wall.

6. Pick any point above the wall that allows you to draw the Fence through both overlapping wall lines. Press the Enter button on the mouse.

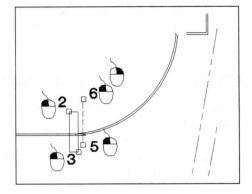

The wall lines should now be trimmed to the end of the glazed portion. Next we will EXTEND the horizontal wall at the top of the curved glazing to its new end location.

1. Pick the EXTEND tool from the Modify tool box.

2. Use a Window to select the end of the horizontal wall, which we moved with the curved glass wall: Pick a point to the left of it and drag the window over the line. Pick a point to the right of the line, as shown in the illustration.

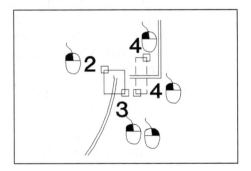

3. Press the Enter button on the mouse to end the Selection Set creation.

4. Pick the two lines of the horizontal wall in turn or use the Fence or Crossing Box method to select the left-hand end of the wall. The wall will be extended by 4 feet to its new endpoint.

Now we're ready to do the serious part of the plan development on paper prior to nailing it down on the CAD system. The issues we should be exploring now are site access, exiting, the structural system, preliminary building core services locations, and interior space allocation. This To Do list is based on what needs to be drawn next on the computer. By the time the client is ready to approve the way the building **looks**, we need to have a pretty good idea of whether or not it **works**.

CHAPTER EIGHT

DRAWING THE STRUCTURAL GRID AND COLUMNS

Contents

We have decided on a steel moment frame structure for this building. A steel frame allows more flexibility in design, and while it is sometimes more expensive than a bearing wall and steel combination, we don't burden the foundation with the weight of a 30-foot plus CMU perimeter wall. We also selected the steel frame system because it is more easily adapted to seismically active geographical regions, and therefore more universally used throughout the country and world.

The column grid should be set to serve our design goals before forwarding it to the structural engineer. Doing this at the earliest stage of schematic design ensures that engineering can begin in parallel, thus avoiding major changes to the drawings at late stages in the project.

The building structure is designed (until engineered) to have wide-flange steel columns and beams. The columns are approximately 28 feet on center, a dimension not dictated by structural calculations but by the footprint of the building. Because **this is not a real building**, we are concerned only with the mechanics of drawing here, not the procedure of setting the structural design. That is something hammered out in real projects between the designer, the engineer, and the client based on a real budget and schedule.

Drawing the Initial Grid Lines

The first structural grid lines are a lot easier to draw than they are to explain and illustrate. I was surprised at the number of screen shots required to avoid confusion, but it seems it goes with the amount of territory covered. Basically, we draw two long lines at right angles to each other and make ARRAYS (multiple copies) of each one in two directions. Believe me, it's not nearly as tedious as doing the same work by hand.

Drawing the First Vertical Grid Line

1. Make Bldgrid the current layer by selecting it from the drop-down layer list, and make sure Ortho Mode is on (press the [F8] key to turn it on, or double-click on the Ortho button on the lower screen button bar).

2. Zoom in on the upper left corner of the building, as shown in the illustration.

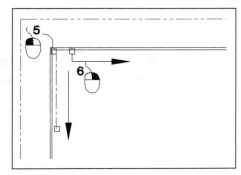

3. Pick the LINE tool from the Draw tool box.

4. Pick the Tracking tool from the Osnap region on the tool bar.

5. Pick the inside upper left wall Intersection for the First Tracking Point. Drag the cursor to the right.

6. Type 1' and [Enter]. Press the Enter button on the mouse to end Tracking. You should now be dragging the grid line down parallel to the east building wall.

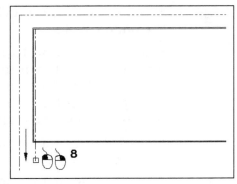

7. Zoom back to the previous view of the building plan.

8. Drag the grid line to the approximate position shown in the illustration. Pick the line's endpoint, then press the Enter button on the mouse to end the Line command.

9. Pick the line again to turn on Grips.

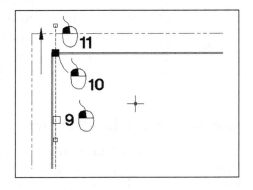

10. Pick the Grip at the top of the line and drag it up above the upper (south) property boundary line.

11. Pick the point for the line to end. Type the CANCEL command [Esc] or [Ctrl]+[C] twice to exit Grips mode.

Draw the First Horizontal Grid Line

Again, make sure that Ortho mode is on, and that you are still on the Bldgrid layer.

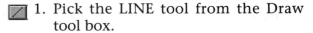 1. Pick the LINE tool from the Draw tool box.

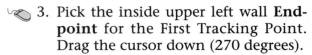 2. Pick the TRACKING tool from the Osnap region on the top tool bar.

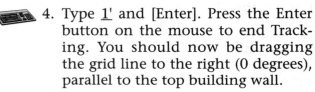 3. Pick the inside upper left wall **Endpoint** for the First Tracking Point. Drag the cursor down (270 degrees).

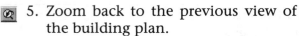 4. Type <u>1'</u> and [Enter]. Press the Enter button on the mouse to end Tracking. You should now be dragging the grid line to the right (0 degrees), parallel to the top building wall.

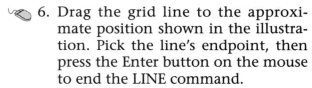 5. Zoom back to the previous view of the building plan.

6. Drag the grid line to the approximate position shown in the illustration. Pick the line's endpoint, then press the Enter button on the mouse to end the LINE command.

7. Pick the line drawn previously to turn on Grips.

8. Pick the Grip at the left end of the line and drag it to the left of the east (leftmost) property boundary line.

9. Pick the point for the line to end. Type the CANCEL command [Esc] or [Ctrl]+[C] twice to exit Grips mode.

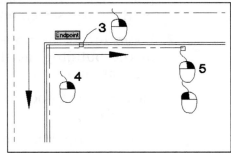

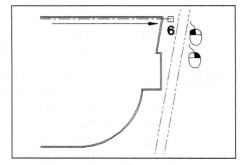

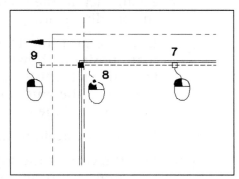

● ● ● ● ● ● ● ● ● ● ● ● ● ● ● ● ● ● ● ●

Create the Horizontal Grid

We now need to place a column grid line 1'-0" from the lower horizontal building wall. We are going to use the DISTANCE tool to find out how to OFFSET the top line to accomplish this.

1. Pick the DISTANCE tool on the tool bar flyout usually signified by the LIST icon.

 2. Pick the Intersection of the two grid lines.

3. Pick the Perpendicular Osnap button and pick the lower inside building wall line. You may want to zoom in and out to make more reliable picks.

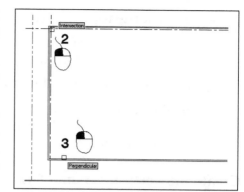

AutoCAD will report that the distance between the upper grid line and the inside of the lower wall is 84'-8". This means we can OFFSET the upper grid line 83'-8" and it will be positioned 1 foot away from the lower inside wall.

 1. Pick the OFFSET tool from the Draw tool box.

 2. Type <u>83'8</u> and press [Enter] or the mouse's Enter key.

 3. Pick the upper grid line.

 4. Pick a point below the upper grid line. The new grid line will appear as shown in the illustration. Press the mouse's Enter key to end the Offset command.

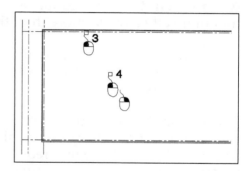

Now we are going to continue using OFFSET to create the two remaining horizontal column grid lines:

 1. Pick the OFFSET tool from the Modify tool box.

 2. Type <u>27'10</u>. This distance will give us two 28'-10" bays between the first grid line and the building walls, and a 28'-0" clear bay in the lateral center of the building.

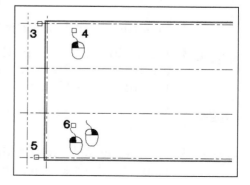

3. Pick the top grid line.

4. Pick a point below the grid line. The new grid line will be created 27'-10" from the original.

5. Pick the lower grid line.

6. Pick a point above the grid line. The horizontal column grid is complete. Press the Enter button on the mouse to end the Offset command.

Create the Vertical Column Grid

We are going to use the ARRAY tool to create multiple copies of the first vertical grid line. The Array command is ideally suited to making quick multiple copies with a uniform distance between the copies. Multiple copies at varying distances apart can be made with the COPY tool, using the Multiple option of the COPY tool, but you will have to manually type in each distance and make many mouse picks, making the ARRAY tool more efficient for more than three or four copies of an object.

1. Pick the Rectangular ARRAY or ARRAY tool from the Modify tool box.

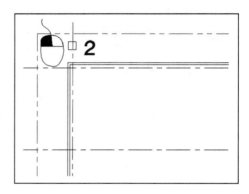

The standard Modify tool box default tool is the two-dimensional ARRAY tool. If you use this, an additional input is required to specify whether you want a rectangular ARRAY or a Polar one. See Appendix A on customizing the tool boxes and tool bar to learn how to install tools such as the Rectangular ARRAY tool.

2. Pick the extended portion of the vertical grid line.

3. Press [Enter] to accept the default <1> number of rows if you are using the 2D Array tool. This step is automatic with the Rectangular ARRAY tool.

4. When prompted for the number of columns, type 8, and press [Enter].

5. Now type 28' for the distance between columns, and press [Enter]. That's all there is to it. We will next add two more ver-

tical grid lines and two horizontal grid lines, and move the last vertical line we just created.

Adding Off-Grid Lines

 1. Zoom in to the right side of the plan, as shown in the illustration.

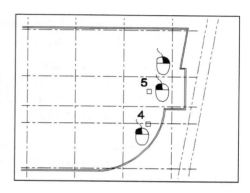

The horizontal grid line near the top of the curved glass wall is 9'-2" below the horizontal building wall at the top of the curve. We want to locate columns at each end of the curve, so we will start by positioning a horizontal grid line 1'-0" above the horizontal wall at the top of the curved glass.

 2. Pick the OFFSET tool from the Modify tool box.

 3. Type <u>10'2</u> and press the [Enter] key.

4. Pick the grid line.

5. Pick a point above the grid line. A new line will be created, as shown in the illustration. Press the Enter button on the mouse to end the Offset command.

The next grid line above our new one is 4'-10" below the short horizontal wall that terminates the lower end of the angled building wall. We are going to locate a column at each corner of the rectangular projection to avoid cantilevering this portion of the structure.

 6. Press the mouse's Enter button to recall the Offset command.

 7. Type <u>3'10</u> and press [Enter] or press the mouse's Enter button.

 8. Pick the grid line.

 9. Pick a point above the grid line. Press the Enter button on the mouse to end the Offset command.

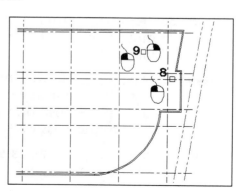

Now we need to make these grid lines shorter so they will not add confusing elements to the drawing:

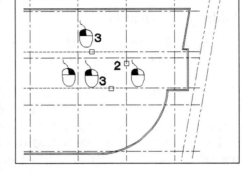

 1. Pick the TRIM tool from the Modify tool box.

 2. Pick the vertical column grid line as shown in the illustration.

 3. Pick the newly created grid lines to the left of the trim line. Press the Enter button on the mouse to end the Trim command.

The rightmost grid line will be used to create our new column position at the top of the curved glass wall:

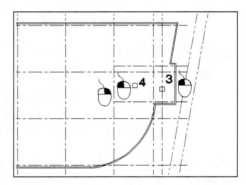

 1. Pick the OFFSET tool from the Modify tool box.

 2. Type 5' followed by [Enter].

 3. Pick the grid line.

 4. Pick a point to the left of the grid line. Press the mouse's Enter button to end the Offset command.

Next we will use the Offset command to create a new vertical grid line at the lower end of the curved glass wall:

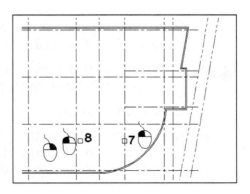

 5. Press the Enter button on the mouse to recall the Offset command.

 6. Type 14'8 followed by the [Enter] key. This distance will put our column in the position we want it relative to the glass wall.

 7. Pick the grid line, as shown in the illustration.

8. Pick a point to the left of the grid line. Press the mouse's Enter button to end the Offset command.

Moving the rightmost column grid line into the desired position is our last step in setting the structural grid design.

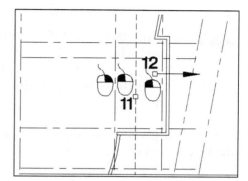

9. The first step is to zoom in to a view like the one in the illustration. Make sure Ortho mode is on.

10. Next pick the MOVE tool from the Modify tool box.

11. Pick the rightmost column grid line. Click the Enter button on the mouse to end the set selection.

12. Pick a point to the right of the line. With Ortho on, drag the line to the right.

13. Type 6'5 and press [Enter]. The line should now be positioned as shown in the illustration.

Now we will use the Trim command and Grips to edit all the grid lines that locate "off grid" columns.

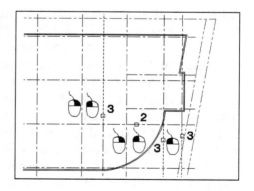

1. Pick the TRIM tool from the Modify tool box.

2. Pick the second grid line from the lower side of the building. Press the Enter button on the mouse to end entity selection.

3. Pick the lines at the positions above and below the trim line, as shown in the illustration. When you have selected all the lines, press the Enter button on the mouse to end the Trim command.

Now turn on the grid line's Grips, and with Ortho on, adjust their length as shown in the illustration.

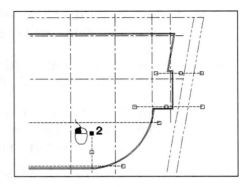

 1. Pick the lines with no command active, to turn on their Grips.

 2. Pick the end grip to activate the Stretch function. Slide the line's endpoint into the positions shown in the illustration.

 3. Enter the CANCEL command twice (press [Esc] or [Ctrl]+[C], depending on your Preferences setting), and then press [R] and [Enter] to redraw the screen.

Our building's structural grid design is now complete. The next step is to put in a series of reasonably sized columns.

Drawing a Steel Column Symbol

We will somewhat arbitrarily pick a wide flange shape of 14 x 211 for tutorial purposes. This represents our schematic design choice anyway; it will be modified when the structural engineer does his or her design work. At that point we will replace the columns with the correct section in a future operation.

A brief word about drawing technique using CAD: Since our drawing is to be plotted at a scale of 1/8" = 1'-0", there is a limit to the fineness of detail that can be produced, and even measured. Our general rule for drawing for this output scale is simple: If you can't measure it on the plotted drawing, don't draw it in detail on the plan. Our columns are prime examples of how this rule works. We need to draw them with some accuracy, because we will want to draw their enclosures to realistic sizes. So it's the overall section size, not the web or flange thickness that will concern us. Added to that will be the fireproofing, and even though it is irregular, we can account for it as an average. What we will draw, then, is a "steel section" that is schematic, intended as a design/dimension guide, not a literal picture of the physical object.

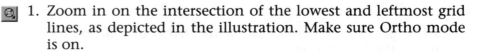 1. Zoom in on the intersection of the lowest and leftmost grid lines, as depicted in the illustration. Make sure Ortho mode is on.

The W14x211 section is 15 3/4" deep by 15 3/4" wide. Allowing for 1 1/4" fireproofing, we get a 17" x 17" overall shape. The web and flanges are indicated with a 2"-wide Polyline.

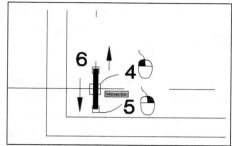

2. Pick the PLINE tool from the Draw tool box.

3. Pick the TRACKING tool from the tool bar.

4. Pick the Intersection of the column grid lines. Set the Polyline Width by typing <u>W</u>, then [Enter] followed by <u>2</u>, and [Enter]. Press the Enter button on the mouse to accept 2" as the ending width.

5. Drag the polyline down (270 degrees), and type <u>8.5</u>, followed by [Enter]. Press the Enter button on the mouse to end Tracking.

6. Drag the polyline up (90 degrees), and type <u>17</u>, followed by [Enter], and press the [Enter] key again to end the Pline command.

That completes the center web of the column. Next comes the top flange, which we will MIRROR to create the bottom flange.

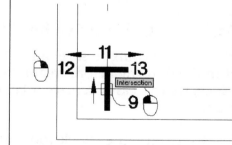

7. Press the Enter button on the mouse to recall the PLINE command. This time we don't need to specify the width, having set it previously. It will stay at 2" until we change it.

8. Pick the Tracking tool from the tool bar.

9. Pick the Intersection of the grid lines again. Drag the pointer up (90 degrees).

10. Type <u>7.5</u> followed by [Enter]. This places our 2"-wide line an inch below the endpoint of the web line.

11. Drag the Polyline to the left and type <u>8.5</u>, and press [Enter].

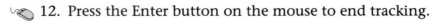
 12. Press the Enter button on the mouse to end tracking.

13. Drag the Polyline to the right and type <u>17</u>, followed by [Enter].

Leave Ortho mode on. We are now going to use the MIRROR command/tool to create the other flange. I use Mirror instead of Copy or Offset because it requires no calculation of offsets. This makes the process easier, and it's just as fast as Offset or Copy. This selection of commands and tools follows my Third Law of CAD drafting: When all other things are equal, pick the command that is the easiest to use.

1. Pick the MIRROR tool from the Modify tool box.

2. Pick the top flange. It will be highlighted, as shown in the illustration. Press the Enter button on the mouse to end the selection process.

3. Pick the **Midpoint** of the web.

4. Drag the mirror line to the right or left and pick any point. Remember, you must have Ortho on. Press the Enter button on the mouse to accept the default <N> to not delete the original object.

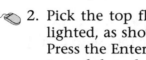

Making a Column Symbol Block

We are now going to save our W14x211 section as a Block, then reinsert it in the drawing before creating the rest of the columns. By making it a block, we can easily replace it later with another Block in a single operation.

1. Pick the Block tool (BMAKE) from the Draw tool box.

2. Type the block name <u>W14x211 column</u> into the dialog box Block Name window.

3. Pick the <u>S</u>elect Objects button and select the column with a window,

by picking the upper left corner first, then the lower right corner, as shown in the illustration. Press the Enter button on the mouse to end the selection process.

4. If the Retain Objects box has a checkmark in it, pick it to cancel this mode so that the column will be erased automatically when the block is created. Pick the Select Point< button.

5. Pick the Midpoint of the column web.

6. Pick the OK button in the dialog box. The dialog box will disappear and so will the column symbol.

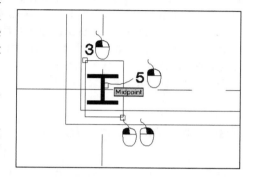

1. Pick the INSERT BLOCK tool on the Draw tool box.

2. Pick the Block button on the dialog box.

3. Pick the W14x211_column Block from the menu. Pick the OK buttons (you can't press the [Enter] key or mouse button; it just works this way—don't ask).

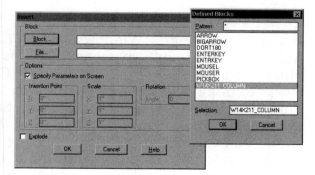

4. Pick the Intersection of the column grid lines as the block insert point. Press the Enter button on the mouse three times to accept the default scales and rotation (0 or none).

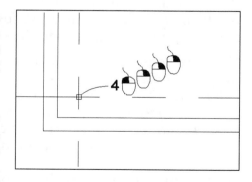

We should now have our column back, only as a block. Blocks are a special type of object in an AutoCAD drawing. Our column block is actually a drawing within our A-1.dwg file, and we can even export it to its own drawing file on disk by using the WBLOCK command.

It does not have to be stored on disk using the same name as the block name, but that is the best practice, as I shall explain in a minute.

Blocks are used to store repeated objects in a drawing file, because they help to keep the file size as small as possible. Instead of treating blocks like an ordinary part of the drawing, with each line endpoint and vector stored in the drawing data, AutoCAD only tracks a block's insert point(s) and stores just one instance of the block geometry, regardless of the number of times a block is repeated. Blocks also make editing the drawing easier, because we can replace many repeated objects at once by substituting one block for another.

· ·

Rules for Naming Blocks

Block names can be thirty-one characters long, and may contain letters, numbers, the $, - (hyphen), and _ (underscore) characters. No spaces or other special characters are allowed. AutoCAD stores all letters in uppercase.

If you are working in Windows 95 or NT, you can take advantage of AutoCAD's thirty-one-character block names when building a disk library of block files (by using the WBLOCK command to copy the blocks to disk). Your file names can be as long as the block name, and can be identical. This means that anyone sharing your block files can see descriptive names such as "Door_rt_hand_90_degrees" used both as file and as block names in the drawing. Long descriptive names are much easier to understand in the block menu. In earlier DOS versions of Auto-CAD, there was no block menu, and block names had to be typed in at the command line each time a block was inserted in the drawing. As a result, many veteran AutoCAD users have formed the habit of using short cryptic abbreviations for block names.

Not only does this limit the number of possible names, but it can create hours of endless searching for the right component or detail to insert in a drawing. Third-party software authors have made lots of money selling add-on programs to organize, track, display, and otherwise manage block libraries. Of course, the easiest, most commonsense thing to do is to use long descriptive block and file names in Win 95/NT!

Naming Blocks Shared with Consultants

One caveat: If you are sharing files with consultants who are using Windows 3.x or DOS to run AutoCAD, you will need to keep names of the blocks and files they will need to access to the eight-character DOS limit. Otherwise Windows 95/NT will create a DOS file name out of eight characters of your thirty-one-character name, and the result could be incomprehensible, for both you and your consultant.

Saving the Block to a File

We are going to finish our Block-making work by writing our column Block out to an individual drawing file, where it will be available for use in other drawings. The first step is to create a directory in which to store our block "library":

1. Minimize AutoCAD to an icon by clicking on the Minimize button on the Windows 95/LT title bar.

2. Open Windows 95 Explorer and create a new folder in your AutoCAD folder (subdirectory) or standalone folder under your main folder (directory). I use a name for the folder such as "Building_Components", then create folders for Electrical, Steel, Plumbing, and other symbols and parts.

3. Close Explorer and click on the AutoCAD task bar button to go back to the drawing.

Now focus your attention on the command line. This is where we will be doing most of our work.

Using the WBLOCK Command

At the Command: prompt, type W+[Enter]. You could type the entire command, but "W" is the standard AutoCAD command alias, stored in the Acltwin.pgp or Acad.pgp file. Unless you have removed or replaced the entry in the .pgp file, "W" should work; otherwise type the WBLOCK command out.

4. The Create Drawing File dialog window will appear (shown below), allowing you to select your directory.

5. In the file name window, type the first Door Block's name, <u>W14x211 _ column</u>, and press [Enter] or pick the OK button. Note: Do not leave a space in the file name or AutoCAD

will not recognize it. Use an underscore or a hyphen instead of spaces.

 6. AutoCAD will now prompt you for the block name. Type <u>W14x211 _ column</u> and press [Enter]. The file will immediately be written to disk, into the directory you created.

Now any drawing can have the column symbol inserted in it just by selecting the File option instead of Block in the Insert Block dialog box. Once a file is inserted into a drawing, all subsequent inserts can be done by picking the Block option and selecting the new block from the Block list.

Creating the Grid of the Building Columns

Make sure your Running Osnap is set to Intersection.

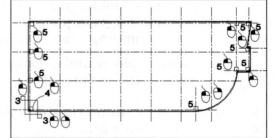

1. Pick the Zoom Previous button on the tool bar, or on the Zoom Flyout, until the screen resembles the illustration.

2. Pick the COPY tool from the Modify tool box.

3. Use a Window to select the column. Remember to pick the leftmost point (corner) of the window first, and to drag the window to the right. Press the Enter button on the mouse to complete the selection process. Type <u>M</u> to select the <u>M</u>any option. Press the mouse's Enter button.

4. Pick the column Block's Insert point using the Insert Osnap tool, by picking any part of the block.

Note: Using the **Insert** Osnap allows us to work zoomed out and to pick the center of the block with precision while not worrying about picking the midpoint of the column web instead of the midpoint of a grid line by accident.

---

5. Now pick the three leftmost grid line Intersections, and the column will be copied to each new location. Next pick the irregular grid intersections we created near the right side of the building. Press the Enter button on the mouse to end the Copy command.

 6. Pick the Rectangular ARRAY tool from the Modify tool box.

 7. Pick the leftmost series of columns using a Window for two pairs of columns, as shown in the illustration. This will ensure that only the columns are picked. Remember to create the windows by picking the leftmost corner first, then dragging the window to the right. Press the Enter button on the mouse to complete the selection process.

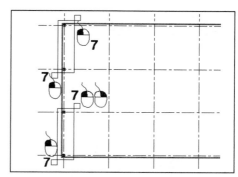

 8. Press [Enter] to accept the default <1> number of rows.

 9. When prompted for the number of columns, type 7, and press [Enter].

 10. Now type 28' for the distance between columns, and press [Enter]. Our column pattern should now be complete.

Note: We could have used the Multiple Insert (MINSERT) command instead of ARRAY, and it would have been slightly faster, but we would not have been able to EXPLODE the blocks in the array created by the Multiple Insert procedure. It's possible that the structural engineer would want to modify our column symbols, so we avoided making them uneditable.

The next step is to locate and draw the doors and windows, including the entrance vestibule. Be sure to save the drawing before going on to the next step.

CHAPTER NINE

Drawing Doors and Creating a Door Library

Contents

135

It may seem strange to devote a chapter to doors, but they are one of the overlooked time burners in CAD drafting, especially in Auto-CAD drawing. It would seem that a door symbol is a simple thing to execute, especially in hand-drafted drawings. On small projects, door symbols can become Works of Art for some drafters, architects, or interior designers. On large projects such as hotels or medical facilities with hundreds or thousands of doors, they are a major pain to draw, even using the most economical of means.

Doors are even more of a pain to create schedules for, even on medium-size commercial projects. The promise (dream) of automated door schedule generation became one of the Holy Grails of AutoCAD drafters, who realized that if doors were inserted into the drawing as blocks, their block definitions could be extracted and counted, and the data generated could be turned into a schedule. Some people had the bright idea that tagging the door blocks with attributes containing finish, door number, and door type information at the time of the block's insertion would provide all the data needed for the door schedule in one fell swoop.

There was one small problem with this last idea. At the time you are drawing the first doors in a building, you have **no idea** what the **final** finish and hardware specification is going to be. Even if you think you know what the final design will require, it is guaranteed to change, often many times and perhaps after the drawings have been issued to bid and for permit. Tracking changes to door finish, type, hardware, etc. over the course of a large project can become a nightmare when the information is "built in" at an early stage.

Dealing with exceptions also becomes more error producing as the project progresses. Drafters would often EXPLODE a door block to create an exception for which no predrawn symbol existed, rather than go through the extra steps of making a new block. The result was an "invisible" door that would never get listed on the "automated" schedule. As a project progressed, door finishes might be split up among different trades under separate contracts as the project was "value engineered." Such a situation can make even a schedule structure meaningless. My general rule on medium- to large-size commercial projects is: **The tighter the budget, the more complex is the door schedule**.

What we are left with is this: In the last ten years, all possible methods of coping with doors in CAD have been tried. One firm might have great success with a certain method on a particular project; a different firm may use the exact same method on another project but suffer abject failure and chaos. I leave the system for data extraction and door scheduling to you. I have seen no solution that works for everyone everywhere in every practice across all project demands, time schedules, and building types. I do know that the more complex the design, the more likely it is that door schedule systems will need to be tediously and repeatedly checked and corrected by knowledgeable people. And we all know what people do: They make mistakes.

● ● ● ● ● ● ● ● ● ● ● ● ●

Drawing a Door

Here we introduce the simplest possible one-line, one-arc door symbol. We will examine the introduction of complexity after this exercise.

Making the Opening for the First Exit Stair Door

The first thing we must do is to locate the starting point for the exit door, which is 2 feet to the left of the column grid line.

ZOOM into a Window covering the area shown in the illustration below. We will use Tracking to draw a short line two feet from the **Intersection** of the grid line and the outside wall, following the numbered steps.

Make sure Ortho Mode is on, and that you have made Bldgwalls the current layer.

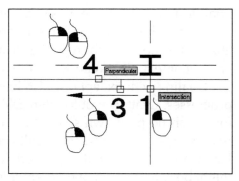

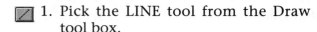 1. Pick the LINE tool from the Draw tool box.

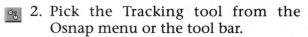 2. Pick the Tracking tool from the Osnap menu or the tool bar.

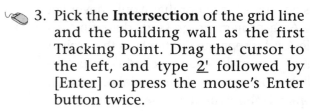 3. Pick the **Intersection** of the grid line and the building wall as the first Tracking Point. Drag the cursor to the left, and type <u>2'</u> followed by [Enter] or press the mouse's Enter button twice.

4. Pick the **Perpendicular** Osnap, and then pick the opposite wall line. Press the Enter button on the mouse to end the LINE command.

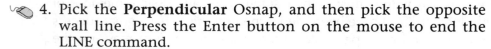 5. Pick the OFFSET tool from the Modify tool box.

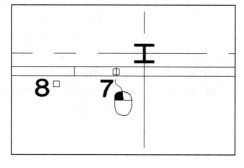 6. Type <u>3'</u> (or <u>36</u>) and press [Enter].

7. Pick the line you just drew.

 8. Pick a point to the left of it.

We would normally TRIM the walls between the jamb lines, but we will wait to do so after we have made the door into a BLOCK symbol.

Yes, there are somewhat faster ways to insert doors and create door openings. We could write a Lisp program or Diesel macro that we could install under a custom icon, which would insert a door and jamb block, activate the trim command, and create the opening while allowing us to parametrically (in Lisp only) pick the direction of door swing and swing angle. I have written and hacked more such routines than I care to remember. However, a funny thing happened when I started bench marking my Lisp programs against the Windows version of AutoCAD LT: My Lisp program required so many user inputs to handle all the variables of door creation that the speed advantage it once had under DOS all but vanished. The introduction of Tracking when used with Direct

Distance Entry is the biggest difference, allowing the setting of the opening position as easily and as quickly as my Lisp code did. The only thing that remains marginally faster is drawing the door swing arc, but by using a block for the door, we can beat or equal that time too.

The Americans with Disabilities Act (ADA) has much to do with this situation. All commercial buildings must have 3'-0"-wide doors from now on. The only exceptions are nonaccessible toilet stalls and fitting rooms (in retail stores). There is no reason to draw a wide variety of door widths unless your work consists entirely of Type 5 (residential) construction. Purely custom Type 5 projects are declining in the practice of architecture and represent the least appropriate application of CAD to architecture, having little reusable components, equipment, and details. Tract homes use standard components common in the industry and can benefit from integrating manufacturers' door and window CAD libraries.

Because commercial construction is competitively bid, at the design development phase we need to work with generic symbols that will support the normal commercial bidding process rather than manufacturers' detailed CAD libraries. Commercial manufacturers of building components realize this and will work to supply detail information on products only when there is a prospect of winning the sourcing for the project. This has resulted in fewer CAD libraries being available for the (admittedly) wider variety of commercial construction products.

Draw the Door Leaf and Swing Arc

We'll draw the door and swing now, make them into a block, and discuss these issues some more.

Make the Bldgdoors layer the current (active) one. Turn Ortho Mode on.

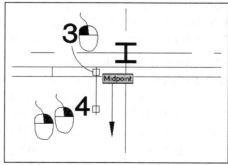

1. Pick the LINE tool from the Draw tool box.

2. Pick the **Midpoint** of the left jamb to start the door leaf.

3. Drag the door line down (270 degrees) and type <u>3'</u>; then press the mouse's Enter button. Press the Enter button again to end the LINE command.

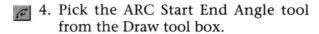 4. Pick the ARC Start End Angle tool from the Draw tool box.

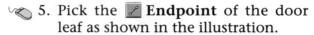 5. Pick the ✎ **Endpoint** of the door leaf as shown in the illustration.

 6. Pick the ✎ **Midpoint** of the right jamb line.

7. Drag the cursor up (90 degrees) and pick a point. Press the Enter button on the mouse to end the Arc command. Voilà, a door swing arc in four mouse picks.

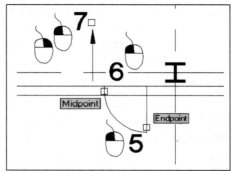

If your firm insists that door swing arcs be on a separate layer so they show as a lighter line weight or line type, make sure to switch to that layer before creating the arc. I think it is easier to just change the color or line type of the door swing arc by selecting the arc, then selecting another color from the color list on the Object Properties tool bar, keeping it on the same layer with the door. Now that we have covered this, we will digress to address some of the religious issues involving doors in building plan drawings.

Religious Issues Concerning Door Symbols

Snyder's First Law of CAD drawing states: **Never put anything in a CAD drawing that you wouldn't put in a paper drawing.**

There are some exceptions for elements you have already drawn, or that are part of CAD libraries available to you, but the basic law is sound. If something is too much trouble to draw relative to the time available in the production budget in manual drafting, we usually don't draw it, or we use an abbreviated symbol for the element.

The same principle holds in CAD drafting, especially if there is no visible or productive difference when the drawing is plotted at the final scale. I have seen architects draw detailed door jambs for every door on a drawing to be plotted at 1/8" = 1'-0". While block insertion makes this relatively efficient, there is no graphic value gained in plotting blobs on door openings (blobs are exactly what are produced on a 1/8" scale plot). Plan graphics aside, the effort is wasted in representing something that is going to be subject to the bidding process and that should be explicit only in large-scale details and specifications, and that should never be part of small-scale plan drawings.

A corollary to this law is: If the graphic element doesn't affect communication with the bidding contractor(s), then don't draw it, or at least don't draw it differently from other graphic elements in the plan. My favorite example is door swings. I know a solid line is supposed to represent a built object, but conventionally everyone understands that door swings are not "real" and therefore don't need to be drawn with a different line type or line weight than the door leaf. Block insertion can make this point moot, but taking the time to create the unique swing arc in the block is still time wasted.

By the same token, I see door leaves of elaborate types created for many popular AutoCAD third-party symbol libraries. Some are drawn using Polylines of defined thickness, and others are actually Rectangles. I have even seen some drawn as rectangular assemblies of single Line segments and even one drawn using the Trace object. For most hinged doors, the most important thing is that they are positioned correctly relative to the jamb so that clearances around them can be determined within an inch. Approach clearances are mandated by federal law and enforced by building department officials and the federal Department of Justice. Get them wrong at your own peril. The value of a door leaf as a graphic symbol is virtually nil. The contractor is going to order doors from the door schedule and install them according to the door number. He will simply ignore the graphic, so don't waste time on it.

On the subject of door numbers: Some third-party architectural front-end software for AutoCAD force you to insert door number keys when inserting door blocks. This is what I call a Murphy System (creating conditions for things to go wrong). Snyder's Second Law of CAD drawing states: **Never draw anything out of the normal sequence of paper drawing development or you will redraw it many times before the final plots are run.**

If you make door number insertion part of your door blocks, you will invite endless editing of drawings all through the construction drawing process because door ratings will change, new finishes will be introduced, and hardware will be revised ad infinitum. The more doors there are on a project, the more changes there will be. Remember that in manual drafting, we knew this and never numbered the doors or made the door schedule until the end of the drawing process. Don't draw foolishly in CAD just because the computer makes it appear easy to do. Even third-party programs that dynamically update the door schedule from a central database can't handle door blocks that have been exploded, or drawings that have been modified off-site and replaced in the network. Don't invite Murphy into the project.

Creating a Door Symbol Block

So, let's make a simple block of our simple door symbol:

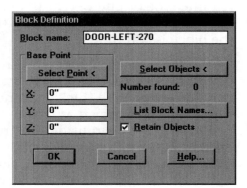

1. Pick the Block (BMAKE) tool from the Draw tool box. You should see the Block Definition dialog box in the illustration.

2. Type the block name in the Block Name window: <u>DOOR-LEFT-270</u>. This name tells us that the block is a door, has the strike side on the left (when we face the inside of the door), and opens down at a right angle to the drawing coordinate system (270 degrees). You can even add the word *degrees* after 270 if you want. (Note: You don't have to type the name as either upper- or lowercase. AutoCAD automatically converts the name to uppercase.) Now pick the **Select Objects** button.

3. Select the door and swing arc with a crossing box by picking the right-most corner first, then the left corner as shown in the illustration. Press the Enter button on the mouse to end the selection process. The dialog box will appear again.

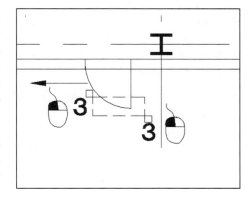

4. If the <u>R</u>etain Objects box has a checkmark in it, pick it to cancel this mode so that the door will be

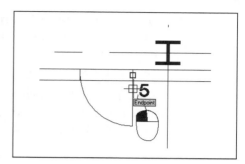

erased automatically when the block is created. Pick the **Select Point <** button.

5. Pick the ✏ **Endpoint** of the door leaf at the jamb end.

6. Pick the OK button in the dialog box. The dialog box will disappear and so will the door symbol.

Inserting a Door Symbol into the Drawing

Now we will INSERT the door to confirm that it was created properly.

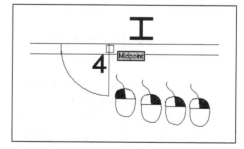

1. Pick the INSERT Block tool on the Draw tool box.

2. Pick the **Block** button on the dialog box.

3. Pick the DOOR-LEFT-270 Block from the menu. Pick the OK buttons.

4. Pick the ✏ **Midpoint** of the right door jamb line as the block insert point. Press the Enter button on the mouse three times to accept the default scales and rotation (0 or none).

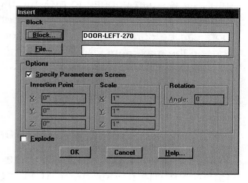

The door should be inserted and appear just as it did when we originally drew it. To quickly check if the insert point is correct, pick the door without issuing any command, thus turning on the block's Grip. There will be only one Grip displayed, because we are working with a block. The Grip should be at the midpoint of the jamb line.

If the door is not properly located, you can quickly correct the problem by grabbing its Grip and moving it to the jamb's midpoint.

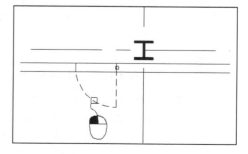

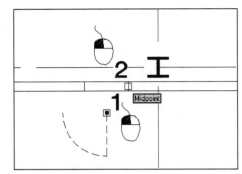

 1. You do this by picking the Grip, which will change color. Pick the Grip again. The door will now follow the cursor.

 2. Pick the ◢ **Midpoint** Osnap tool from the tool bar, and then pick the door jamb line.

Note: Even if you are fairly well zoomed out and your pick box covers both the jamb line and the wall lines, AutoCAD is smart enough to pick the most likely Midpoint, which will always be the shortest line, which is of course our door jamb. This means that inserting and copying doors can be done at a fairly high zoom factor (low magnification), allowing you to work on the maximum visible section of the plan.

● ●
Creating the Second Exit Door

We are now simply going to use COPY to create the door for the exit stair from the upper floor. It is located 2'-0" to the right of the column. Make sure Ortho Mode is on.

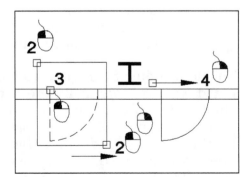

1. Pick the COPY tool from the Modify tool box.

 2. Select the door and jambs with a Window, drawn by picking points from left to right, as shown in the illustration. Press the Enter button on the mouse after picking the second window point, to end the selection process.

 3. Pick the **Intersection** of the leftmost jamb and wall as shown in the illustration.

4. With Ortho on, drag the door and jamb to the right. Type 7' and press the mouse's Enter button or the [Enter] key. Seven feet is the width of the door (3'-0") plus the 4'-0" the doors will be separated (2 feet to each side of the column grid line).

Now all we need to do is TRIM the walls at the door openings.

1. Pick the TRIM tool from the Modify tool box.

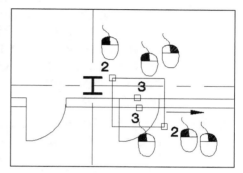

2. Select the trimming objects by drawing a Window around the door jambs only, as shown in the illustration. Press the Enter button on the mouse after picking the second window point, to end the selection process.

3. Pick the wall lines between the jambs. Press the mouse's Enter button to end the Trim command.

This is a perfect example of when and how to get the most out of COPY and Direct Distance Entry. We could have duplicated the steps in creating a new door opening and inserting our Door Block, but picking the right tool at the right time can further decrease the effort needed.

There is no equivalent to this process in manual drafting, where the tools used don't vary that much. A CAD drafter needs to be constantly thinking ahead to the next step, thinking of tools and processes that best fit the task. Naturally, that results in CAD's imposing a higher workload on you, as compared with manual drafting, because you have to think not only about the drawing but about the mechanics of drawing it. There's no muscle memory (kinesthetic retention) to substitute manual skill for memory in CAD: You are constantly thinking before doing. Those with the best thinking process will draw the best. One purpose of this book is to highlight that thinking process so that you can be prepared to look at the drawing as a series of problems to be solved, where each of them is slightly different each time.

Making a Second Set of Exit Doors

Note: You may want to use the ✎ **Match Properties** tool to change column grid lines 5 and B to the Bldgwalls layer temporarily, to make their intersections with the building wall lines easier to pick. To do this, pick the Tool button, pick any wall line, then pick the two grid lines (B is second from the top, 5 is fifth from the leftmost), and press the mouse's Enter key.

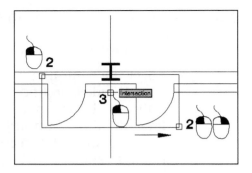

1. Pick the COPY tool from the Modify tool box.

2. Select the doors and jambs with a Window, drawn by picking points from left to right, as shown in the illustration. Press the Enter button on the mouse after picking the second window point, to end the selection process.

3. Pick the **Intersection** of the building grid line and the outer side of the building wall, as shown in the illustration, as the point to COPY **from**.

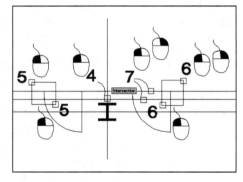

4. Pick the ZOOM Previous tool, or if that doesn't work, the ZOOM Extents tool, then ZOOM Window onto the fifth column from the left at the upper (south) wall of the building so that it is magnified as shown in the illustration. Drag the doors and their connecting wall into position and pick the **Intersection** of the grid line and the **inside** of the building wall as the point to COPY **to**.

5. Pick the TRIM tool on the Modify tool box. Select the leftmost door jamb line with a Window.

6. Select the rightmost jamb line with a window. Press the Enter button on the mouse to complete the selection.

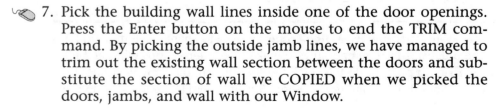

7. Pick the building wall lines inside one of the door openings. Press the Enter button on the mouse to end the TRIM command. By picking the outside jamb lines, we have managed to trim out the existing wall section between the doors and substitute the section of wall we COPIED when we picked the doors, jambs, and wall with our Window.

Reverse the Direction of the Door Swing

We now need to reverse the doors so they swing out from the building wall. This is easily done with the MIRROR tool and requires a minimum number of mouse picks.

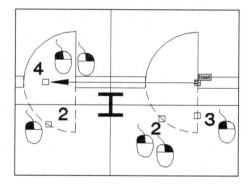

1. Pick the MIRROR tool from the Modify tool box.

2. Make sure Ortho Mode is on. You may want to turn Onsap Mode off by double clicking the Osnap button on the bar below the drawing screen. Pick the doors (since they are blocks, one pick gets both door and swing).

3. Pick the 🖲 **Insertion Point** Osnap button. Pick one of the doors.

4. Drag the mirror line toward the opposite door and pick a point in space.

5. Type Y followed by [Enter] to erase the old doors and end the MIRROR command.

Make the Third Pair of Exit Doors

We are going to do this by COPYing the doors we just finished to a new location on the east (left) wall, ROTATING them into position, and then TRIMming away the unneeded portions of the east wall, just as we did in the last operation.

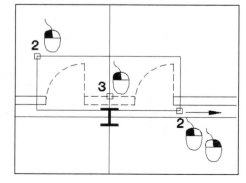

1. Pick the COPY tool from the Modify tool box.

2. Select the doors, jambs, and wall between the doors with a Window, as shown in the illustration. Press the Enter button on the mouse to complete the selection process.

3. Pick the **Intersection** of the outside building wall and grid line as the

COPY **from** point. Pick the ZOOM Previous or ZOOM Extents button so you can see the entire plan. ZOOM in on the region of column grid lines 1 and B (the top horizontal line is A, the leftmost line is 1) on the east wall, as shown in the illustration.

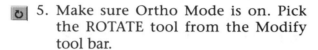

4. Pick the **Intersection** of the B grid line and the east exterior building wall as the COPY *to* point. The doors and wall section should appear, inserted at a right angle to the east wall, as shown.

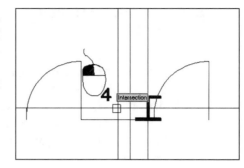

Rotate the Copied Door Assembly into Position and Trim the Walls

5. Make sure Ortho Mode is on. Pick the ROTATE tool from the Modify tool bar.

6. Use a Window to select the new copy of the doors, jambs, and wall. There's no way to avoid selecting the column as well, but not to worry.

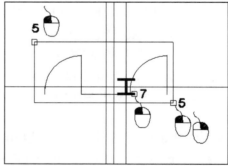

7. Hold down the [Shift] key and pick the column block, which will remove the column from the selection set.

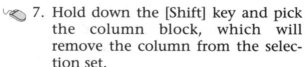

8. Now pick the **Intersection** of the outside wall line and the horizontal grid line, as shown in the illustration.

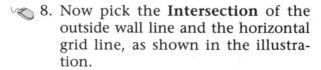

9. Drag the mouse pointer up (90 degrees) to rotate the door assembly into position. Pick a point in space free from an Osnap attachment.

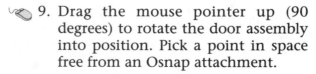

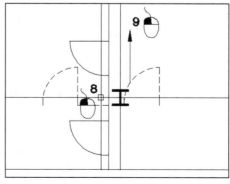

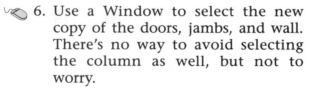

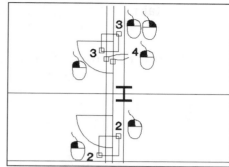

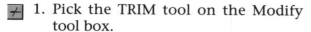 1. Pick the TRIM tool on the Modify tool box.

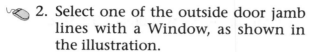 2. Select one of the outside door jamb lines with a Window, as shown in the illustration.

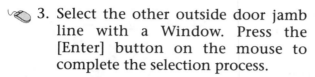 3. Select the other outside door jamb line with a Window. Press the [Enter] button on the mouse to complete the selection process.

4. Pick the wall lines inside one of the door openings to trim out the east wall portion we don't want. Our exit door insertion is complete.

If you previously changed the properties of grid lines 5 and B, change them back to match the rest of the grid using the 🔲 Match Properties tool.

Drawing the Recessed Entry Vestibule

Set the current layer to Bldglazing. 🔲 ZOOM Extents to see the whole building, or dynamically 🔲 PAN to see the whole curved glazing area at the northeast corner of the building.

Place the Entry in the Curved Wall

We are going to create a recessed entry centered on the midpoint of the curved curtain wall. The methods we will use work just as well for an off-center location for the recess of the vestibule. This is a good, generalized method for creating any design element in a curved surface.

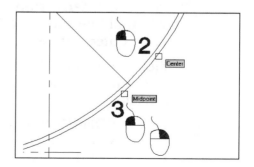

1. Pick the LINE tool from the Draw tool box.

2. Pick the **Center** Osnap button, then the outside arc of the curved curtain wall as the starting point for a construction line.

3. Pick the **Midpoint** Osnap button, then the outer arc of the curved cur-

tain wall again. Press the Enter button on the mouse to end the LINE command.

With the center construction line now complete, we will create the side walls of the 8'-0" vestibule:

 4. Pick the OFFSET tool from the Modify tool box. Type <u>4</u>' followed by [Enter].

 5. Pick the center construction line.

 6. Pick a point above the construction line.

 7. Pick a point below the center construction line. Press the mouse's Enter button to end the OFFSET command.

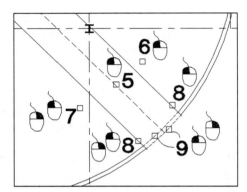

 8. Pick the TRIM tool from the Modify tool box. Select the side wall lines. Press the mouse's Enter button to complete the selection process.

 9. Select each curved curtain wall line between the side walls. Press the Enter button on the mouse to end the TRIM command.

To draw the back of the vestibule, we could employ a variety of methods, but this one is the simplest by far, out of the six or so I have tried:

 1. Pick the LINE tool on the Draw tool box.

 2. Pick the **Intersection** of the upper side wall line and the curved curtain wall's outer surface.

 3. Pick the **Intersection** of the lower side wall line and the curtain wall's outside line.

 4. Pick the OFFSET tool on the Modify tool box. Type <u>4</u>' and [Enter] to set the OFFSET distance.

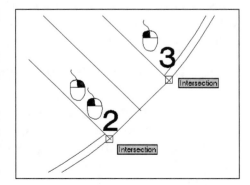

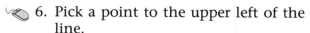 5. Pick the line we just drew.

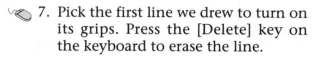 6. Pick a point to the upper left of the line.

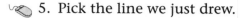 7. Pick the first line we drew to turn on its grips. Press the [Delete] key on the keyboard to erase the line.

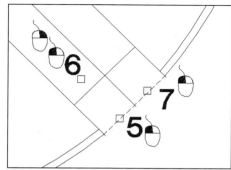

Erasing objects this way is faster than using the ERASE command, because it requires no [Enter] key press to confirm the erasure. You can select as many objects as you want before pressing [Delete].

Now we are going to create the outer glass walls of the vestibule recess by using OFFSET and FILLET:

 1. Pick the OFFSET tool on the Modify tool box.

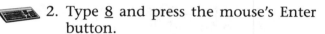 2. Type <u>8</u> and press the mouse's Enter button.

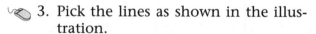 3. Pick the lines as shown in the illustration.

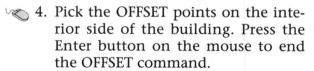 4. Pick the OFFSET points on the interior side of the building. Press the Enter button on the mouse to end the OFFSET command.

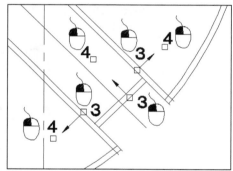

Now for the FILLET work. Make sure the FILLET Radius is set to *zero*.

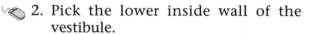 1. Pick the FILLET tool on the Modify tool box.

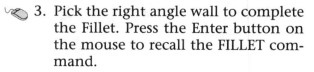 2. Pick the lower inside wall of the vestibule.

3. Pick the right angle wall to complete the Fillet. Press the Enter button on the mouse to recall the FILLET command.

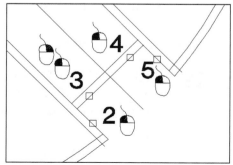

4. Pick the same line again, near the top corner of the vestibule.

 5. Pick the upper side wall of the vestibule to complete the second Fillet.

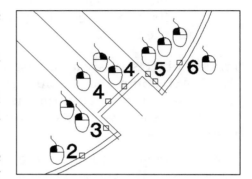

 1. Press the Enter button on the mouse to recall the FILLET command.

 2. Pick the lower inside curved curtain wall line.

 3. Pick the lower inside vestibule wall line. Press the mouse's Enter button to recall the FILLET command.

 4. Pick the inside back wall of the vestibule. Press the mouse's Enter button to recall the FILLET command. Pick the back wall of the vestibule again.

 5. Pick the upper inside wall of the vestibule. Press the mouse's Enter button to recall the FILLET command. Pick the same wall line again.

 6. Pick the upper inside curved curtain wall line to complete the last FILLET.

• •

Inserting the Double Entry Doors in the Vestibule

Now we'll make a door opening and insert into it the DOOR-LEFT-270 Block we created earlier, then use the MIRROR tool/command to make the second door.

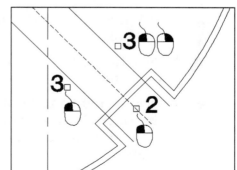

Make the Door Opening

 1. Pick the OFFSET tool on the Modify tool box. Type 3' and press [Enter] on the keyboard or mouse.

 2. Pick the center construction line as the object to OFFSET.

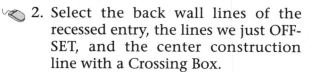 3. Pick points to either side of the center construction line. Press the mouse's Enter button to end the OFFSET command.

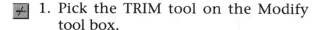 1. Pick the TRIM tool on the Modify tool box.

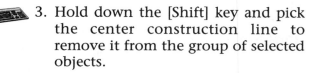 2. Select the back wall lines of the recessed entry, the lines we just OFF-SET, and the center construction line with a Crossing Box.

3. Hold down the [Shift] key and pick the center construction line to remove it from the group of selected objects.

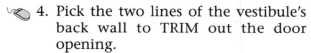 4. Pick the two lines of the vestibule's back wall to TRIM out the door opening.

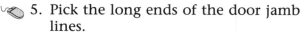 5. Pick the long ends of the door jamb lines.

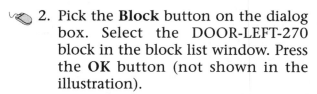 6. Pick the other ends of the door jamb lines. Press the mouse's Enter button to end the TRIM command.

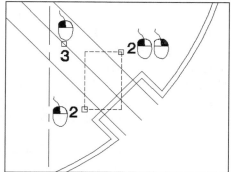

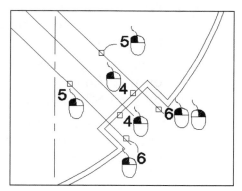

Insert the Doors

1. Pick the INSERT Block tool on the Draw tool box.

2. Pick the **Block** button on the dialog box. Select the DOOR-LEFT-270 block in the block list window. Press the **OK** button (not shown in the illustration).

3. Pick the 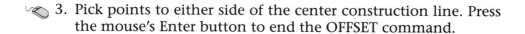 **Midpoint** Osnap button on the Osnap tool box. Pick the right-hand jamb line as shown in the illustration. Press the mouse's Enter button twice to accept the default scale factors for the door.

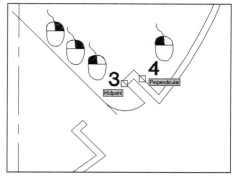

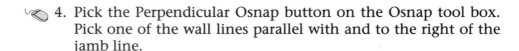

🖱 4. Pick the Perpendicular Osnap button on the Osnap tool box. Pick one of the wall lines parallel with and to the right of the jamb line.

The final step is to mirror this door to create the second of the pair. In Release 12, MIRRORed blocks could not be EXPLODEd for manipulation of their parts, and as a result, we got used to EXPLODING blocks before MIRRORing them. Since Release 13, that has not been necessary, but you should remember the situation with Release 12 if you are sharing drawings with consultants who use Release 12, including blocks that were not uniformly SCALEd or MIRRORed.

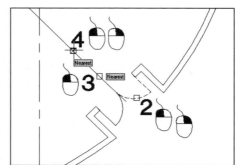

🔲 1. Pick the MIRROR tool in the Modify tool box.

🖱 2. Pick the door. Press the mouse's Enter button to complete the selection process.

🖱 3. Pick the ⬛ **Nearest** Osnap button on the Osnap tool box. Pick a point on the center construction line.

🖱 4. Pick the **Nearest** Osnap button again, and pick a point farther up the center construction line. Press the mouse's Enter button to end the MIRROR command.

• • • • • • • • • • • • • • • • • • •

Making a Door Library

The following illustration shows the file called Door_Library.dwg, which is included on the companion disk for this book. It provides all the basic configurations of doors needed to draw any building or interior design. All these doors are 3'-0" in width, as required by the federal ADA law. You are free to use them by creating blocks of your own from this drawing. The insert points are indicated by the solid round dots. The block names in the illustra-

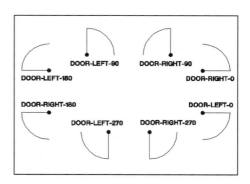

tion are suggestions. Right and left refer to the side of the door that is hinged, as seen from the inside.

When you make your door blocks, don't forget to write them out to their own files on the hard disk, so they will be available from all your drawings.

Make a Variable Size Door Block

If you do a lot of residential work or other drawings with doors that vary from the 3'-0" width standard, you may find it advantageous to create a variable door block, which allows you to specify the width when it is inserted. Here's how:

1. Make a COPY of one of the doors in the drawing.

2. Pick the SCALE tool on the Modify tool box.

3. Use a Crossing Box to select the leaf and swing of the door. Pick a point near the door as the base point.

4. When prompted for the scale factor, type .027777 and press [Enter]. Our 36"-wide door will now be a 1"-wide door.

Zoom in close enough to see it clearly, and make it into a block. Write it out to disk so it will be accessible from other drawings, using the WBLOCK command.

Now, when you insert this 1" x 1" door, specify its size in inches by entering the size as the block's scale factors. For instance, we would make it a 2'-6"-wide door by typing 30, [Enter], and [Enter] when prompted for the inserting *x* and *y* scale factors.

GLAZING SYSTEM PLAN DRAWINGS

Contents

The following is a simplified method for creating glass and mullions in plan drawings that are meant to be plotted at small scales. The idea is not to create a complex drawing at the detail level. That type of detail is best copied from glazing systems manufacturers' literature, some of which is even available in CAD format.

• •

Drawing the Mullion System in Plan View

Our tutorial building uses Kawneer's FA-SET 400 butt glazing system with backing mullions as the beginning design concept. The system may be changed later, but it promises to meet the midrange budget for such a project and to provide the appearance of butt-joined glass from the exterior.

Drawing the Mullion System Construction Lines

Turn Ortho Mode on. Make Bldglazing the current layer. Zoom in to the view shown in the illustration below.

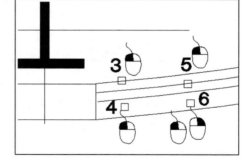

 1. Pick the Offset tool from the Modify tool box.

 2. Type 2 and press the Enter button on the mouse.

 3. Pick the back glazing line.

 4. Pick a point below the line, as shown in the illustration.

 5. Pick the new offset line (our new construction line).

6. Pick a point below it, and then press the Enter button on the mouse to end the Offset command. This new line is the line of the glass itself.

Drawing the First Mullion

 1. Pick the Polyline (PLINE) tool from the Draw tool box. Pick the Tracking tool from the Draw tool box.

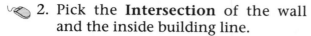

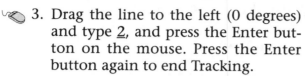

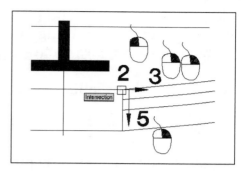

2. Pick the **Intersection** of the wall and the inside building line.

3. Drag the line to the left (0 degrees) and type <u>2</u>, and press the Enter button on the mouse. Press the Enter button again to end Tracking.

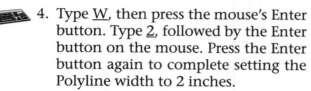

4. Type <u>W</u>, then press the mouse's Enter button. Type <u>2</u>, followed by the Enter button on the mouse. Press the Enter button again to complete setting the Polyline width to 2 inches.

5. Drag the line down (270 degrees) and type <u>4</u>; then press the mouse's Enter button (yes, one more time).

Making the Mullion Block

We now have a 2" by 4" mullion symbol. To make it truly useful, we need to take one more step and turn it into a block.

Note: You may want to zoom in a bit to make picking the small Polyline easier, or to temporarily resize your pick box.

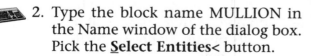

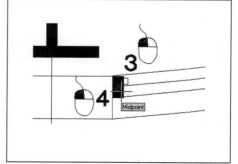

1. Pick the Make_Block tool (BMAKE) from the Modify tool box.

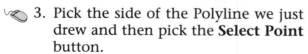

2. Type the block name MULLION in the Name window of the dialog box. Pick the **<u>S</u>elect Entities<** button.

3. Pick the side of the Polyline we just drew and then pick the **Select Point** button.

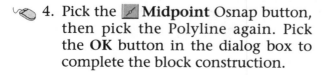

4. Pick the **Midpoint** Osnap button, then pick the Polyline again. Pick the **OK** button in the dialog box to complete the block construction.

Now for the fun part of using the block: We are going to insert it in the curved glazing system using one command and a few keyboard inputs, without using complex math to figure out the on-center dimensions.

Inserting Mullions in the Curved Curtain Wall

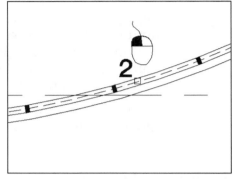

 1. Pick the DIVIDE tool from the Modify tool box.

 2. You are now prompted to select the object to divide. Pick the construction line as shown in the illustration.

 3. The next prompt asks for the number of segments or a block name. Type [B] and press [Enter] to use our Mullion block.

4. When prompted for the block name, type MULLION and press [Enter]

5. AutoCAD will ask if you want to align the block with the object (the curved construction line). Press [Enter] to accept the default answer <Y> (Yes).

6. Finally you will be prompted for the number of segments to divide the construction line into. Type 6 and press [Enter]. Press [Enter] again to end the Divide command.

As you can see, mullions have appeared at approximately 4'-0" on center along the arc of the construction line. In truth, it took two tries to hit on this spacing, but AutoCAD allows us to do quick studies of several possibilities and to get rid of unsatisfactory results by simply pressing [U]ndo. We decided to go with six segments because they give us an almost true 4'-0" window width, with some space left over to allow for sealant, expansion, and construction tolerances.

If you use Divide to insert objects in a line, circle, or arc, the objects must be blocks. This requires an extra step, but making a block is trivial compared with trying to calculate the width of six equal segments of a partial circumference of a circle.

Inserting the Corner Frames

Now let's complete the framing around the entry recess. Zoom in to a view close to that shown in the illustration. Use the OFFSET tool to Offset the glazing and construction lines as we previously did for the curved lines.

1. Pick the Insert Block (INSERT) tool from the Draw tool box. Pick the Block button, then the MULLION block from the block list. Pick OK.

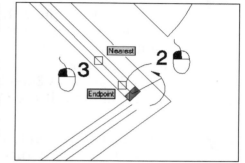

2. Pick the **Endpoint** Osnap button, then pick the lower portion of the construction line as shown in the illustration. Press the Enter button on the mouse twice to accept the default *x* and *y* scale factors.

3. Pick the **Nearest** Osnap button, then rotate the block into position and pick the middle construction line again.

Insert the next corner frame by following the same steps as the previous ones:

1. Press the Enter button on the mouse to recall the INSERT command. Pick the OK button to use the MULLION block (still shown in the Block name window) again.

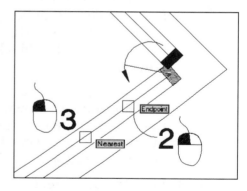

2. Pick the **Endpoint** Osnap button, then pick the lower portion of the curved construction line as shown in the illustration. Press the Enter button on the mouse twice to accept the default *x* and *y* scale factors.

3. Pick the Nearest Osnap button, then rotate the block into position and pick the middle construction line again.

Now we will complete the framing at the inner corner and door jamb:

Inserting the Inner Corner and Jamb Frames

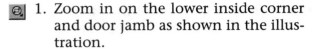

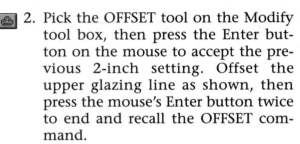

 1. Zoom in on the lower inside corner and door jamb as shown in the illustration.

2. Pick the OFFSET tool on the Modify tool box, then press the Enter button on the mouse to accept the previous 2-inch setting. Offset the upper glazing line as shown, then press the mouse's Enter button twice to end and recall the OFFSET command.

3. Type 1, then [Enter] to reset the Offset distance, and offset the door jamb line to the lower left as shown.

Inserting the Inner Corner Framing

Make sure that the Running Osnap is set to Intersection.

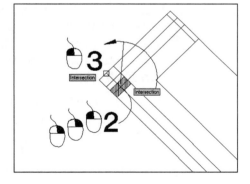

1. Pick the Insert_Block (INSERT) tool from the Draw tool box. Pick the OK button to use the MULLION block (still shown in the Block name window) again.

2. Pick the **Intersection** of the construction line as shown in the illustration. Press the Enter button on the mouse twice to accept the default *x* and *y* scale factors.

3. Pick the **Intersection** of the construction and frame lines, then rotate the block into position and pick the middle construction line again.

Repeat the steps for the next mullion:

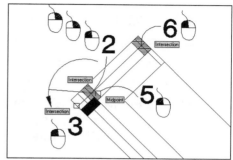

1. Press the mouse's Enter button to recall the INSERT command. Pick the OK button on the Insert dialog box.

2. Pick the Intersection of the construction line and curtain wall frame as illustrated. Press the mouse's Enter button twice to accept the default *x* and *y* scale values.

3. Pick the Intersection of the end of the construction line and inside curtain wall frame.

4. Pick the Copy tool on the Modify tool box. Pick the mullion we just inserted. Press the Enter button on the mouse to complete the selection.

5. Press the Midpoint Osnap button, then pick the mullion again.

6. Pick the Intersection of the construction line and the line you previously OFFSET one inch.

Creating the Curtain Wall Framing on the Other Side of the Doors

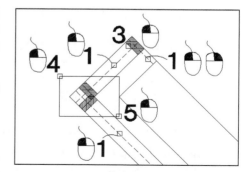

1. Pick the construction lines as shown in the illustration. Press the [Delete] key on the keyboard to ERASE the lines.

2. Pick the MIRROR tool from the Modify tool bar.

3. Pick the jamb frame.

4. Pick the inside corner frame with a window as shown in the illustration.

5. ZOOM out (ZOOM Previous) as shown.

6. Pick the glass lines.

7. Select the corner framing with a window as shown in the illustration.

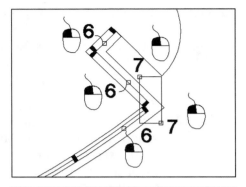

8. ZOOM out so you can see the entrance recess and lower curved glass wall completely.

9. Pick the lower section of the mullions with a Window as shown in the illustration. ZOOM in temporarily if you need to. Make sure to include the mullion where the glass wall and building wall meet.

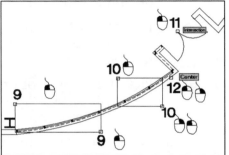

10. Pick the next group of mullions with another Window as illustrated; press the Enter button on the mouse to end the selection process.

11. Pick the Intersection of the door arcs as the start point of the mirror line.

12. Pick the Center Osnap button. Pick the outside line of the curved wall framing, as shown in the illustration. This effectively creates a mirror line from the arc's center to the center of the wall, at the double door intersection. Press the Enter button on the mouse twice (once to end the command, again to accept the default to not delete the objects being mirrored), and all the mullions and corner frames will be precisely duplicated on the other section of the curved wall.

As a last task, Zoom in and erase the last curved construction line.

We have now placed all the framing and glazing on the curved curtain wall. Placing framing in the other windows of the building will wait until we have designed their shape and locations in elevation drawings.

In working through the design development on the CAD system, we make decisions in a different order than is customary when developing design with hand-drawn drawings. At this time, we are using AutoCAD's precision to help us nail down the geometry of the building's plan, and once that is done to a certain point, we will finish the design by using the plan to generate the elevations. We will finish design development on the elevations and then reflect that geometry back to the plan(s) to complete them.

CHAPTER ELEVEN

DRAWING DESIGN ELEVATIONS

Contents

We will now use the design elevation(s) to take advantage of CAD's full-size drawing abilities to set the basic building geometry for all work going forward. Our level of drawing detail will remain schematic. We are more interested in where things stop and start than in the impact of specific building systems on every minute inch of the plan and elevation geometry.

It is here that we begin to use the design elevation drawing to provide the foundation for construction drawings and to determine with some precision the design constraints for the structural engineer. Our tutorial building is simple enough that we can resolve most of these issues by drawing the design of the north wall only. More complex projects will need to repeat this process several times. On large projects, we would pick the "linchpin" areas of the design and move to the less demanding "generic" areas just short of starting final construction drawings.

The process illustrated here is that of projecting the plan to the elevation, studying design elements in elevation, and projecting the elevation design back to the plan. The idea is to lock down the design geometry at this point so that it can be the basis for everything to follow, once the client has approved the design.

• •

Beginning the Elevation Layout

ZOOM All to see the view of the entire plan shown in the following illustration. The elevation of the North wall will be drawn directly below the lower side of the plan (in the parking lot area). Make sure the Site-boundary layer is *off* and that Ortho Mode is on.

Establish the Parapet Line

 1. Pick the LINE tool on the Draw tool box.

 2. Pick a point to the left of the lower building corner. Drag the line to the right (0 degrees).

 3. Pick the line's endpoint past the right side of the building. Press the Enter button on the mouse to end the Line command.

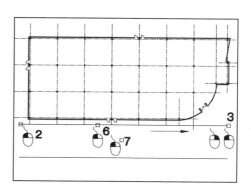

Establish the Floor Line

  4. Pick the OFFSET tool from the Modify tool box.

5. Type <u>35'</u> and press [Enter], for the Offset Distance. As you will recall from the 3D study we did earlier, in Chapter 7, our floor to floor and floor to roof deck distance is 16'-0", giving us a slab to roof height of 32'-0". Thirty-five feet allows us a 3'-0" parapet, thus the offset distance of floor to parapet of 35 feet.

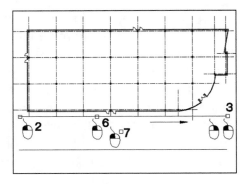

6. Pick the parapet line.

7. Pick a point below the line.

Project Wall and Column Grid Lines

 1. Pick the EXTEND tool from the Trim/Extend Flyout on the Modify tool box.

2. Pick the new floor line. Press the Enter button on the mouse to end the selection.

3. Pick the lower portion of the left-most exterior wall line. If you need to ZOOM in to do it, go ahead, then ZOOM Previous to see the whole plan again.

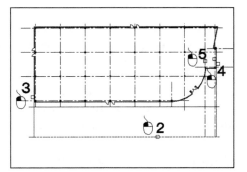

4. Pick the rightmost outside wall line in the plan, as indicated in the illustration.

5. Pick the two right column grid lines, as shown in the illustration. At any time, if you pick the wrong line, just press the [U] key to Undo the selection. When all the lines are selected, press the mouse's Enter button to Extend them.

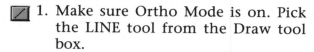

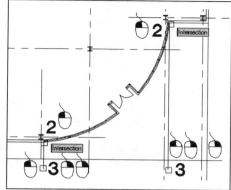

1. Make sure Ortho Mode is on. Pick the LINE tool from the Draw tool box.

2. Pick the ⊠ **Intersection** of the building wall and the curved curtain wall line, or the ▱ **Endpoint** of the curved curtain wall. Drag the line below the elevation's parapet line.

3. Pick a point below the parapet line and press the Enter button on the mouse to end the line. Press the mouse's Enter button again to recall the Line command and draw the other line as shown in the illustration.

Project the Curtain Wall Mullion Centers

Again, make sure Ortho Mode is on. Set the running Osnap to Endpoint. Zoom in to the view shown in the next illustration.

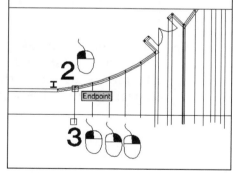

1. Pick the LINE tool or press the mouse's Enter button to recall the Line command.

2. Select the **Endpoints** of the mullions in the plan, and drag lines below the parapet line in the elevation. Zoom in to pick points if necessary, Zooming out to complete the lines.

3. Pick points below the parapet line to end each line. Press the Enter button on the mouse to end and recall the Line command each time, until all the lines are drawn as illustrated. Also draw lines from the intersection of the door swing arcs.

 1. Pick the EXTEND tool on the Modify tool box.

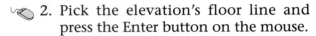 2. Pick the elevation's floor line and press the Enter button on the mouse.

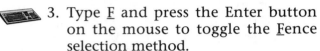 3. Type F and press the Enter button on the mouse to toggle the Fence selection method.

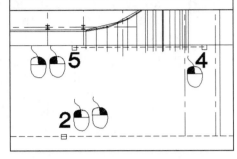

4. Pick a start point for the Fence line as illustrated at the right.

5. Pick an end point for the Fence, and press the mouse's Enter button to Extend the lines.

Zoom in to see the column grid lines better, as shown in the next illustration:

1. Pick the BREAK-1 point tool from the Modify tool box.

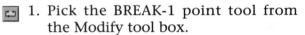 2. Pick a point on the column lines approximately where shown in the illustration. This will be the first point where the grid line will be broken.

3. Pick a second break point, as illustrated, and press the Enter button on the mouse to recall the command. Repeat for the other grid lines.

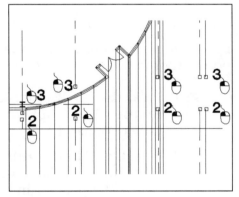

1. Pick the TRIM tool on the Modify tool box.

2. Pick the parapet line. Press the Enter button on the mouse to end the selection process.

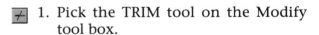 3. Again, type F and press the Enter button on the mouse to toggle the Fence selection method.

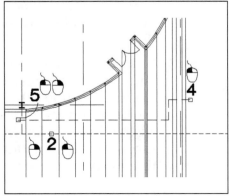

4. Pick a start point for the Fence line above the parapet line.

 5. Pick an end point for the Fence, and press the mouse's Enter button to Trim the lines above the parapet.

Trimming the Entry Vestibule and Doors

 1. Pick the OFFSET tool from the Modify tool bar.

 2. Type <u>10'</u> and press the Enter button on the mouse.

 3. Pick the floor line.

 4. Pick a point above the floor line to create the top line of the entry recess.

 5. Select the TRIM tool in the Modify tool box.

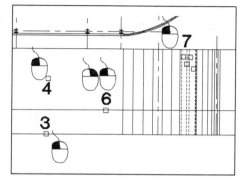

 6. Pick the line we just created and press the Enter button on the mouse to complete the selection process.

 7. Pick the lines indicated in the illustration; make sure to trim only the first two lines (counting from the left) of the outside corner frame on the right-hand side of the entry. Press the Enter button on the mouse to end the Trim command.

Using a Schematic Elevation to Generate Part of the Roof Plan

Now we are going to draw one more elevation line, then go back to the plan to set the projection of the atrium roof. The elevation line we are going to draw is the angle of the glass curtain wall as it slopes up to the roof over the clerestory windows of the atrium. This line will set the plan location of the edge of the roof, which we will draw before doing any more work on the elevation.

This is one example of how CAD drafting and design can be more productive than manual methods. On the CAD system, we can work with elevations and plans on the same virtual "sheet," using elevations to provide design elements for plans, and vice versa. Alignment and geometric accuracy are easily maintained. When we are finished, we will be 90 percent of the way to final construction drawings. Changes at this stage are easy to handle because all the drawings are working together, and the precision of the CAD system allows changes to be made in what will become the final drawing, not some design intent drawing that must be redrafted to create a construction document.

Draw the Curtain Wall Slope from Floor to Roof

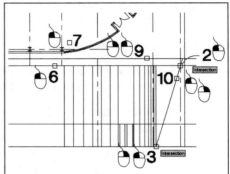

1. Pick the LINE tool from the Draw tool box.

2. Start the line at the ☒ **Intersection** of the column grid line and the parapet line, as indicated in the illustration.

3. End the line at the ☒ **Intersection** of the floor line and the last glass line on the right side of the curtain wall. Press the Enter button on the mouse to end the line command.

4. Select the OFFSET tool on the Modify tool box.

5. Type <u>3'</u>, followed by [Enter].

6. Pick the parapet line. We are going to create the top of the curtain wall's projection above the parapet, which will also give us 3-foot-high clerestory over the lobby atrium.

7. Pick a point above the parapet line. This Offset is also the curtain wall roof line.

8. Select the EXTEND tool from the Modify tool box.

9. Pick the new curtain wall roof line, then press the mouse's Enter button to end the selection process.

10. Pick the upper end of the line we drew in steps 1 to 3 to EXTEND it to the roof line. Press the mouse's Enter button to end the Extend Command.

In the next operation, we will use this point to establish the farthest projection of the inclined curtain wall at the roof level.

Create the Roof Plan of the Atrium

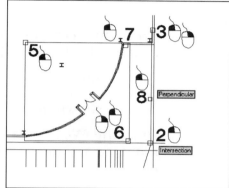 Zoom out and then window back in to the view shown in the next illustration. Turn Ortho Mode on, and turn the Bldgrid layer off.

1. Select the LINE tool from the Draw tool box.

2. Pick the **☒ Intersection** of the angled construction line and the roof line as the starting point.

3. Drag the line up (90 degrees) to the building plan, and pick a point to end it.

4. Pick the COPY tool from the Modify tool box.

5. Pick a starting point for a selection window, just above the end of the curved curtain wall, and just to the left of the column at the bottom of the curve.

6. Drag the selection window down to the right to select the entire curtain wall and entrance recess. Press the Enter button on the mouse to end the selection process.

7. Pick the **Endpoint** Osnap button, then the upper end of the outside arc of the curtain wall, as illustrated here. Zoom in if you need to in order to make a sure pick.

8. Pick the **Perpendicular** Osnap button, then pick the vertical construction line we drew earlier. You should now have a copy of the curtain wall with one end attached to the upper horizontal building wall, and the other floating out in space. The next task is to quickly remove the entry doors and recess, and add a mullion in the center of the curve.

Next we will erase the entry recess and doors but will preserve one side of the entry glazing to FILLET to the upper half of the curtain wall.

Erase the Duplicated Entry

 1. Type E (ERASE), and press the mouse's Enter button, or pick the ERASE tool on the Modify tool box.

 2. Pick a starting point for the first selection Window, as illustrated.

 3. Stretch the Window over the upper door and recess wall, then pick the corner point for the Window.

 4. Start another selection Window at a point completely above the lower door, as shown in the illustration.

 5. Drag the Window around the door, including the corner frame, jamb, and sidelight. Pick the lower corner of the window.

 6. Start another window just big enough to surround the outside corner frame of the lower curtain wall.

 7. Complete the window by picking its lower corner. Click the Enter button on the mouse to finish the erasure.

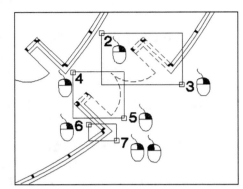

Note: An easy way to ERASE without issuing a command is to use the Crossing Box or Window to select the objects to ERASE. This will turn on their Grips, which can be ignored. Once all objects are selected, simply press [Delete] on the keyboard to execute the erasure.

Complete the Curtain Wall at the Roof

 1. Pick the FILLET tool on the Modify tool box. Check the fillet radius shown on the command line. If it is not zero, type R followed by [Enter] and then type 0, and [Enter]. Press [Enter] again to recall the FILLET command.

 2. Pick the center glass line on the remaining part of the entry recess.

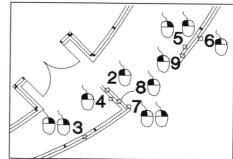

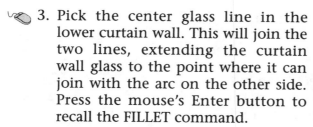 3. Pick the center glass line in the lower curtain wall. This will join the two lines, extending the curtain wall glass to the point where it can join with the arc on the other side. Press the mouse's Enter button to recall the FILLET command.

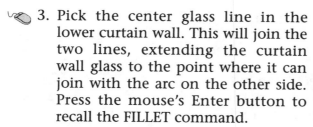 4. Next pick the inside vertical line on the remaining recess wall.

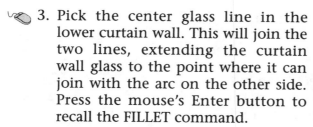 5. Now pick the inside line of the upper curtain wall. Press the mouse's Enter button to recall the FILLET command.

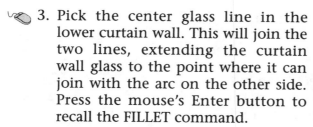 6. Pick the outside line of the upper curtain wall section.

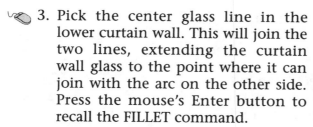 7. Pick the old exterior side of the vestibule glazing, as shown in the illustration. Press the mouse's Enter button to recall the FILLET command (one last time).

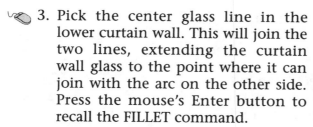 8. Pick the center glass line in the remaining part of the recess.

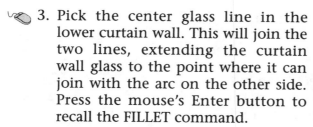 9. Finally, pick the center glass line of the upper section of curtain wall.

All the arcs should now be joined at their endpoints. The next step is to erase the remaining parts of the recessed entry and to put a center mullion in the curtain wall above the entry.

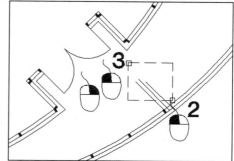

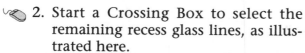 1. Press the E key on the keyboard and the Enter button on the mouse or pick the ERASE tool on the Modify tool box, or use the [Delete] key method explained earlier.

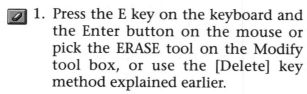 2. Start a Crossing Box to select the remaining recess glass lines, as illustrated here.

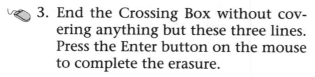 3. End the Crossing Box without covering anything but these three lines. Press the Enter button on the mouse to complete the erasure.

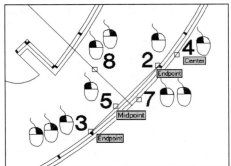

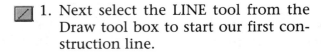 1. Next select the LINE tool from the Draw tool box to start our first construction line.

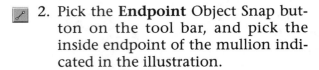 2. Pick the **Endpoint** Object Snap button on the tool bar, and pick the inside endpoint of the mullion indicated in the illustration.

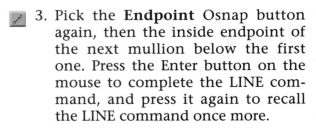 3. Pick the **Endpoint** Osnap button again, then the inside endpoint of the next mullion below the first one. Press the Enter button on the mouse to complete the LINE command, and press it again to recall the LINE command once more.

4. Pick the **Center** point Osnap button, then the outer arc of the curtain wall. This will start our next construction line from the arc's center.

5. Pick the **Midpoint** Osnap button on the tool bar, then pick our first construction line. Press the mouse's Enter button to end the Line command.

6. Pick the EXTEND tool from the Trim/Extend flyout on the Modify tool box.

7. Pick the outer arc of the curtain wall. Press the Enter button on the mouse to end the selection process.

8. Pick the second construction line. Press the mouse's Enter button to conclude the Extend command.

Add a Mullion to the Roof Curtain Wall

Now we're ready to use the Intersection Osnap to COPY a mullion from the first-floor entry to the second-floor curtain wall plan. We need a mullion that is oriented perpendicular to the center of the arc of the curtain wall.

1. Press the <u>E</u> key on the keyboard, then the mouse's Enter button, or pick the ERASE tool on the Modify tool box.

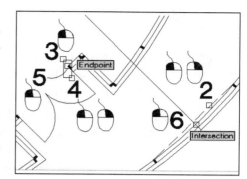

2. Pick the straight construction line. Press the mouse's Enter button to complete the Erase command.

3. Select the COPY command on the Modify tool box, and start a selection Window near the door jamb on the first-floor entry plan, as illustrated here.

4. Be careful to window only the door jamb, leaving the end of the curtain wall return out of the window. Press the Enter button on the mouse to complete the selection.

5. Pick the **Endpoint** Osnap button on the tool bar, then the inside end of the jamb.

6. Pick the **Intersection** of the construction line and the inside line of the curtain wall. The mullion is now installed on the orthographic centerline of the curved curtain wall.

• •

Completing the Elevation of the Curved Curtain Wall Glazing

Now that we have the roof line curtain wall located, we need to project the mullion centerlines onto the elevation so that the glass joint lines will have the proper angle of slope.

Project the Top Locations of the Curtain Wall Mullion Centers

Set the 🔲 Running Osnap to **Endpoint** (uncheck Intersection and any other Osnap selections in the Running Object Snap dialog box), and make sure Bldglazing is the current layer. Zoom out far enough to see the complete curtain wall arc in the plan, and the line of the curtain wall roof in the elevation, as illustrated below.

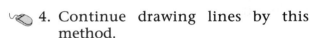 1. Select the LINE tool from the Draw tool box.

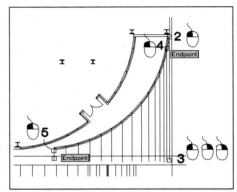

2. Pick the **Endpoint** of the uppermost mullion, and drag the line down (270 degrees).

3. Pick a point below the clerestory roof line, to end the line. Press the Enter button on the mouse once to end the Line command, and again to recall it.

4. Continue drawing lines by this method.

5. At the last line, where the curtain wall joins the building wall, pick the **Endpoint** of the curtain wall arc, instead of the mullion, as the start point.

Trim the Projected Lines at the Top of the Curtain Wall

1. Select the TRIM tool on the Modify tool box's Trim/Extend flyout.

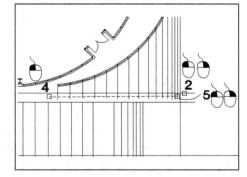

2. Pick the curtain wall roof line, then press the mouse's Enter button to end the selection process.

3. Type F, and [Enter], to use the fence selection method, or pick the Fence button on the Selection flyout in the tool bar.

4. Pick the starting point for the Fence line, as indicated in the illustration.

5. Pick the endpoint for the Fence line, being careful not to cross the line of the outside building wall (Zoom in as necessary to make an accurate pick). Press the Enter button on the mouse to execute the Trim command, then press it again to end the command.

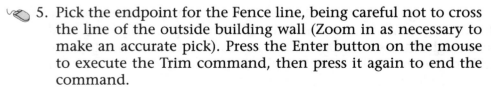

Stretch the Glass Lines to the Roof Line

Set the 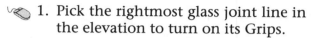 Running Object Snap to **Intersection**.

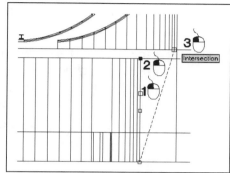

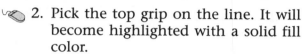 1. Pick the rightmost glass joint line in the elevation to turn on its Grips.

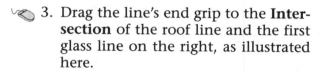 2. Pick the top grip on the line. It will become highlighted with a solid fill color.

3. Drag the line's end grip to the **Intersection** of the roof line and the first glass line on the right, as illustrated here.

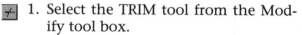 4. Repeat the above procedure until the right eight lines have been stretched into position, as shown in the next illustration.

Trim the Doors to the Entry Recess

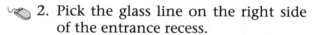 1. Select the TRIM tool from the Modify tool box.

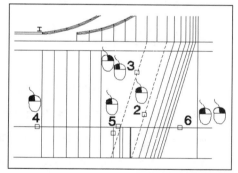

2. Pick the glass line on the right side of the entrance recess.

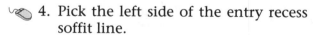 3. Pick the glass line on the left side of the recess. Press the Enter button on the mouse to complete the Trim operation.

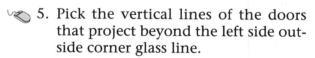 4. Pick the left side of the entry recess soffit line.

5. Pick the vertical lines of the doors that project beyond the left side outside corner glass line.

6. Pick the opposite end of the entry recess soffit line to trim it at the recess opening. Press the Enter button on the mouse to end the Trim command.

Draw a New Glass Joint above the Doors

Check to make sure Bldglazing is the current layer.

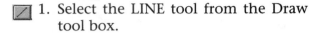

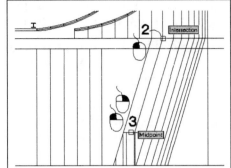

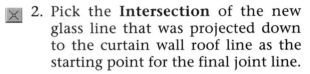

1. Select the LINE tool from the Draw tool box.

2. Pick the **Intersection** of the new glass line that was projected down to the curtain wall roof line as the starting point for the final joint line.

3. Pick the **Midpoint** Osnap button, then the entry soffit line. Press the Enter button on the mouse to end the Line command.

4. Continue to stretch the glass joint lines on the curtain wall to their respective positions at the roof line, until they look like the lines in the next illustration.

Trim the Building and Curtain Walls to the Final Shape

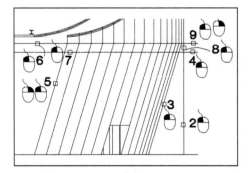

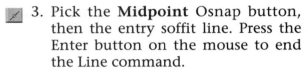

1. Select the TRIM tool from the Modify tool box.

2. Pick the right side building line as the first cutting line.

3. Pick the right side glass line.

4. Pick the building parapet line.

5. Pick the left side glass/wall joint line. Press the mouse's Enter button to complete the selection of cutting edges.

6. Pick the curtain wall roof line to the left of the leftmost glass line, as shown in the illustration.

7. Pick the parapet line to trim it out of the curtain wall by selecting a point between the outside glass edges.

8. Pick both extensions of the right side building wall to trim it at the corner.

9. Pick the right end of the curtain wall roof line where it extends past the curtain wall's right edge. Press the Enter button on the mouse to complete the Trim command.

Remove the Projected Curtain Wall Glass Joint Lines

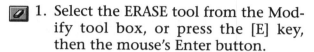

1. Select the ERASE tool from the Modify tool box, or press the [E] key, then the mouse's Enter button.

2. Type <u>F</u> and [Enter] to use the fence selection method, or pick the Fence tool from the Select flyout.

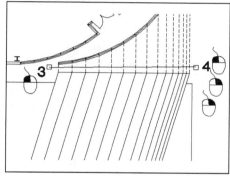

3. Pick the start point for the fence, and drag the fence line all the way through the projected glass joint lines, as illustrated here.

4. Pick the Endpoint of the Fence and press the mouse's Enter button to Trim the lines. Press the Enter button again to end the Trim command.

Create the Intermediate Horizontal Curtain Wall Members

1. Select the OFFSET tool from the Modify tool box.

2. Pick the floor line.

3. Type <u>16'</u> and press the Enter button on the mouse.

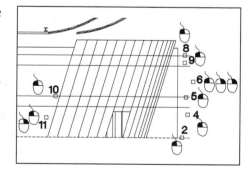

4. Pick a point above the floor line. This will create the line of the horizontal member that aligns with the second floor slab.

5. Pick the new floor/horizontal member line.

6. Pick a point above the line to create the horizontal member that aligns with the roof line at the parapet. Press the Enter button on the mouse to end the Offset command. Press the

Enter button again to recall the Offset command.

 7. Type <u>4'</u> and press the mouse's Enter button.

 8. Pick the horizontal member line we just created.

 9. Pick a point below the line.

 10. Pick the second floor/horizontal member line.

 11. Pick a point below it. Press the Enter button on the mouse to end the Offset command.

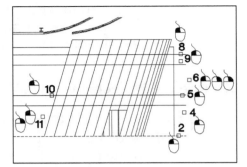

1. Select the TRIM tool from the Trim/Extend flyout on the Modify tool box.

2. Pick the right side of the curtain wall. Press the Enter button on the mouse to end the selection process. Pick the Fence button on the Selection Options flyout on the tool bar.

 3. Pick a point below the lowest horizontal member to start the Fence line.

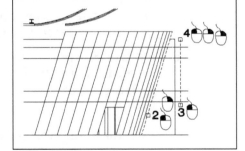

It doesn't matter whether you start above the highest line or below the lowest, but since we had to start somewhere, we picked this point.

 4. Drag the line up above the top horizontal member and pick its endpoint. Press the Enter button on the mouse to trim the lines. Press it again to end the Trim command.

- -

Designing the North Side Window Walls

We are next going to use the angle of the curved curtain wall to create the design for the rest of the building's fenestration. This is a lazy way out, of course, but we are just demonstrating how to use repetitive design elements in CAD drawings. If I were to take it a step further, we could begin to rotate these repeating elements as they work their way around the building's elevation, and to do other interesting things with them. Within the constraints of time and the construction budget, AutoCAD provides tools for playing endless games with a building's geometry.

Remember, this is not a book on design. This book shows how to **draw** design using AutoCAD, and what parts of it to draw at what point in the drawing development process, so your work can be more productive. We spend time refining the plan until we figure out the geometry of the curtain wall at the atrium, and we do not draw the other windows in plan view until we do our final shaping on the elevations in the CAD drawing. It does not matter how many rough or finished sketches I have drawn by hand by this point; what matters is how I spend my CAD hours.

Draw the First Lines for the North Side Windows

Our glazing design places two story curtain walls between column pairs, flanking the exit stair. We are going to hold the vertical sides of these glazed areas one foot to either side of the flanking column center lines. We are going to use the OFFSET command to place these lines and the EXTEND to put them in the elevation.

 Zoom out to the drawing's Extents. You can zoom back in a bit as long as you can see the column grid lines shown in the illustrations.

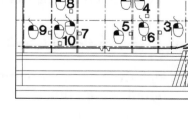

 1. Select the OFFSET tool from the Modify tool box.

 2. Type <u>1'</u> and [Enter] to set the Offset distance.

 3. Pick the right column grid line, as shown in the illustration.

4. Pick a point to the left of the grid line, offsetting it in that direction. Press the mouse's Enter button to recall the OFFSET command.

5. Pick the next column line to the left.

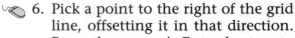

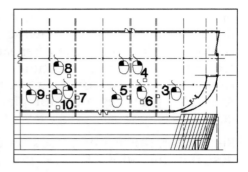

6. Pick a point to the right of the grid line, offsetting it in that direction. Press the mouse's Enter button to recall the OFFSET command.

7. Move to the left pair of column lines and pick the rightmost one.

8. Pick a point to the left of the grid line, offsetting the line in that direction. Press the mouse's Enter button to recall the OFFSET command.

9. Pick the next column line to the left.

10. Pick a point to the right of the grid line, offsetting the line a foot.

That looks like a lot of steps to go through, but it happens very quickly, as fast as you can move the mouse and press its buttons. Now we'll extend the lines.

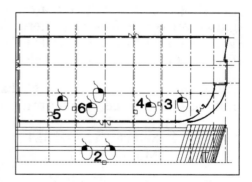

1. Select the EXTEND tool from the Modify tool box.

2. Pick the floor line in the elevation. Press the Enter button on the mouse to finish selecting objects.

3. Pick the lines you just Offset, as shown in steps 4, 5, and 6 in the illustration. After picking the last line, press the Enter button on the mouse to end the Extend command.

Change the Lines to the Bldgwall Layer

1. Pick the Match Properties tool on the tool bar. Pick one of the building wall lines.

2. Type P and [Enter] to recall the Previously selected lines.

3. Press the mouse's Enter button.

The lines will now be on the correct layer and have the Continuous line type.

While we didn't need to change all the lines, it was easier to use the Previous selection method at this time—it saved trying to select just two with the mouse. Now Zoom in to the view shown in the next illustration, and we'll borrow the angle of the curved curtain wall for our glazing design. Make sure you have the Running Osnap set to Intersection.

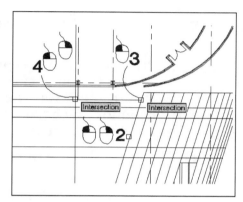

1. Select the COPY tool from the Modify tool box's Duplicate Objects flyout.

2. Pick the leftmost edge of the curved curtain wall. Press the Enter button on the mouse to finish the selection process.

3. Pick the **Intersection** of the parapet line and the curtain wall edge as the first point of displacement.

4. Pick the Intersection of the right construction line and the parapet. Press the mouse's Enter button to end the Copy command.

Trim the Window Wall Edge Lines

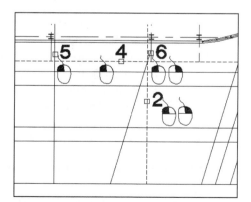

1. Select the ERASE tool on the Modify tool box.

2. Pick the right side construction line. Press the mouse's Enter button to complete the Erase command.

3. Select the TRIM tool from the Modify tool box.

4. Pick the parapet line as the cut line. Press the Enter button on the mouse to end selecting cut lines.

5. Pick the left side window edge above the parapet.

6. Pick the right side (angled) window edge above the parapet. Press the Enter button on the mouse to end the TRIM command.

Array the Window Edge to Create the Glass Joints

 1. Pick the ARRAY tool from the Modify tool box.

 2. Pick the left-hand side of the window.

 3. Press [Enter] to accept the default <1> for the number of rows. Type 6 and [Enter] for the number of columns. Type 4'6 and [Enter] for the distance between columns.

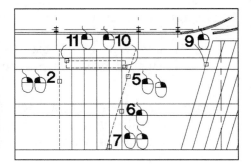

 4. Pick the TRIM tool from the Modify tool box.

 5. Pick the angled side of the window as the trimming edge. Press the mouse's Enter button to finish the selection.

 6. Pick the end of the rightmost glass joint that extends beyond the window edge.

7. Pick the end of the other glass line near the bottom of the window. Press the Enter button on the mouse to end the TRIM command.

Note: We used the ARRAY command to create the glass lines because it saved us some mouse picks where OFFSET would have required marginally more time. In general, if I have to copy an object more than four times in a uniform direction, I use the ARRAY command because I can type fairly quickly. If you have difficulty using the keyboard, OFFSET may still be faster for you.

Change the layer of the new glass lines.

 8. Pick the Match Properties tool on the tool bar.

 9. Pick one of the objects drawn previously on the Bldglazing layer, as illustrated.

🖱 10. Use a Crossing Box to pick the new glass lines.

🖱 11. Press the mouse's Enter button and the properties of the selected lines will be changed to match those of the first object picked.

Having to explicitly change properties of objects is the CAD drafter's unique burden. When we draw by hand, we just make the glass lines the proper line weight and forget about them. In CAD drawing, we must keep the computer informed of our intent to have the glass layer plot at a lighter line weight than the walls. Now for something that's easier to do on the computer than on paper. Zoom out to the view shown in the next illustration. Keep the Running Osnap set to **Intersection**.

Copy the New Window-Wall to the Next Location

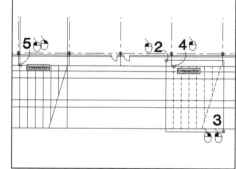

🖱 1. Select the COPY tool from the Modify tool box.

🖱 2. Pick the starting corner of a selection Window as shown in the illustration.

🖱 3. Pick the end of the selection Window below the lower right-hand side of the window-wall, making sure to include the entire top portion. Press the mouse's Enter button to conclude the selection process.

🖱 4. Pick the upper left corner (**Intersection**) of the window as the first point of displacement.

🖱 5. Pick the **Intersection** of the leftmost construction line and the parapet. Press the mouse's Enter button to complete the Copy command.

After taking a look at the glazing design, we have decided we don't like the effect of the angled window walls being copies of each other. We also want to equalize the amount of glass on the two floors. What we want is for the second window wall to be a double mirror image of the first one. We could use the MIRROR tool to mirror the window vertically and then horizontally, but we would have to draw a construction line horizontally at the vertical midpoint of the window wall. Here's an easier method that still takes two steps but doesn't require a construction line:

Rotate the Window Wall to Create a New Design

Zoom in to the view of the second window as shown in the illustration. Make sure Ortho mode is on.

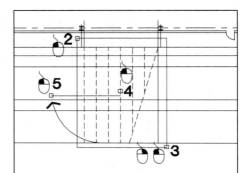

1. Select the ROTATE tool from the Modify tool box.

2. Pick a point to the upper left of the window wall, and drag a selection Window down and to the right.

3. Pick the endpoint of the selection Window and press the Enter button on the mouse to complete the selection.

4. Pick the **Midpoint** Osnap button, then the third vertical line from the left side of the window wall.

5. Drag the cursor to the left (180 degrees) and pick a point. The window wall should now be upside-down and backward.

Put the Left Window Wall into its Final Position

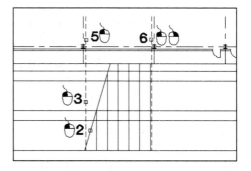

1. Select the MOVE tool on the Modify tool box. Type P followed by [Enter] to recall the previously selected items.

2. Pick the **Endpoint** Osnap button on the tool bar, then pick the lower portion of the left side window wall line.

3. Pick the **Endpoint** Osnap button again, then pick the left side construction line.

4. Select the ERASE tool from the Modify tool box, or use the [Delete] key method to avoid this step.

5. Pick one of the construction lines.

6. Pick the other construction line and press the mouse's Enter button or press [Delete] to complete the erasure.

Completing the Basic Elevation

To complete our elevation we need to add the ground line and extend the building walls down to it. We also need to draw the exit doors. These are the final operations before we tackle exit/access ramp and stair design.

Determine the Elevation of the Ground Line

Before we can draw the ground line in the elevation, we need to find out where it is, relative to the floor line. We know the site slopes a total of 3'-0" from south to north, so we will draw some construction lines to quickly let AutoCAD measure the distance for us. Turn on the Site-boundry layer. Make sure Ortho Mode is on and that the Running Osnap is set to **Intersection**. ZOOM out to the view shown in the illustration.

1. Select the LINE tool from the Draw menu. Pick the upper right **Intersec-tion** of the property boundary lines.

2. Drag the line down (270 degrees) and pick the Perpendicular Osnap button on the tool bar. Pick the lower site boundary line.

3. Drag the line to the right, type 3', and then press the mouse's Enter button.

4. Pick the upper right **Intersection** of the property boundary lines again, and press the Enter button on the mouse to end the Line command. Press it again to recall the LINE command.

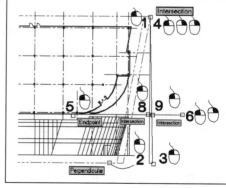

5. Pick the **Endpoint** Osnap button on the tool bar. Pick the endpoint of the lower outside building wall on the plan, to start a construction line. Zoom in if you need to to pick the line accurately.

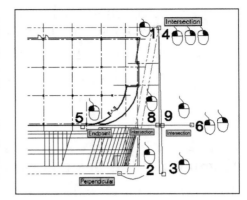

 6. Drag the construction line across the lines we just drew and pick an endpoint. Press the Enter button on the mouse to end the Line command.

 7. Pick the DISTance tool on the tool bar's Inquiry/List flyout.

 8. Pick one **Intersection** of the construction line and the vertical ground/floor lines.

9. Pick the other **Intersection** of the construction line and the ground/floor lines. AutoCAD should now show you the distance in the command line window at the bottom of the drawing screen. It will say:

Distance = 2'-1/16". Angle in XY Plane = 0. Angle from XY Plane = 0.

Delta X = 2'-1/16", Delta Y = 0'-0", Delta Z = 0'-0"

We will ignore the 1/16" and just use 2'-0", since the actual site probably varies a lot more than that.

Project the Exit Doors from Plan to Elevation

Make sure the Bldgdoors layer is the current one, and that the Running Osnap is set to Intersection.

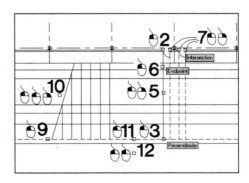

 1. Select the LINE tool from the Draw tool box.

 2. Pick the **Intersection** of the leftmost door jamb, and drag the line down (270 degrees).

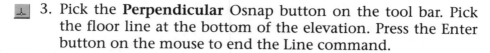
3. Pick the **Perpendicular** Osnap button on the tool bar. Pick the floor line at the bottom of the elevation. Press the Enter button on the mouse to end the Line command.

4. Select the COPY tool from the Modify tool box.

5. Pick the line we just drew. Press the [M] key on the keyboard, followed by [Enter] to select the Multiple Copy option.

6. Pick the **Endpoint** Osnap button on the tool bar. Pick the upper end of the line being copied.

7. Pick the **Intersections** of the walls and the other three jambs of the exit doors, using Running Osnap. Press the Enter button on the mouse to end the Copy command.

Now create the head line of the doors:

8. Select the OFFSET tool from the Modify tool box.

9. Type 7' and press [Enter]. Pick the floor line.

10. Pick a point above the floor line. Press the Enter button on the mouse, once to end the Offset command, and once more to recall it.

While we are still working in the OFFSET command, create the ground line:

11. Type 2' and press [Enter]. Pick the floor line again.

12. Pick a point below the floor line. Press the [Enter] key to end the Offset command.

Complete the Elevation of the Exit Doors

1. Elect the TRIM tool from the Modify tool box.

2. Pick the start point for a Crossing Box as shown in the illustration. Drag the box up and to the left.

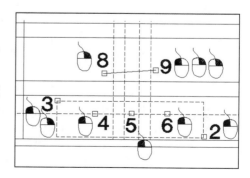

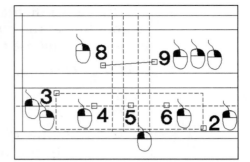

 3. Pick the Crossing Box's endpoint above the door header line and to the left of the doors. Press the Enter button on the mouse to complete the selection of objects.

 4. Pick one side of the header line to the right or left of the doors.

 5. Pick the header line between the doors.

 6. Pick the other side of the header line beyond the doors.

 7. Type F and [Enter] or pick the Fence button on the Selection Options flyout on the tool bar to use the Fence selection method.

 8. Pick the first line of the Fence to one side of the door opening lines, above the headers.

 9. Drag the Fence line through the vertical door lines and pick an end point. Press the Enter button on the mouse to trim the lines. Press it again to end the Trim command.

10. As a final step, use the **Match Properties** tool on the tool bar to change the door headers to the Bldgdoors layer. Pick the Match Properties tool, then one of the doors in the plan, followed by the headers. Press the mouse's Enter button to complete the change.

Trim the Horizontal Curtain Wall Members

1. Select the TRIM tool from the Trim/Extend flyout on the Modify tool box. Pick all the window wall side lines, the left end building wall line on the elevation, and the left side of the curved curtain wall. Last, pick the outside building wall line on the plan and press the mouse's Enter button to end the selection process, as shown in the illustrationon the following page.

2. Pick the Fence button on the tool bar's Select Options flyout, or press the [F] key on the keyboard, followed by [Enter], to use the Fence selection method. Draw Fence lines between the window walls as shown in the illustration.

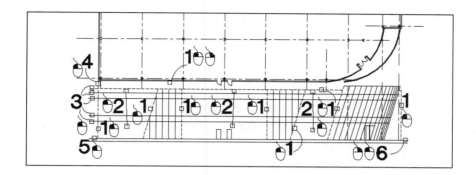

3. Pick the horizontal member and parapet lines extending past the left side of the building.

4. Pick the left building wall line between the plan and the parapet.

5. Pick the left floor line where it extends past the left side of the building wall in the elevation.

6. Pick the right floor line where it extends past the right building wall. Press the Enter button on the mouse to complete the trim operation.

Extend the Building Walls to the Ground Line

1. Select the EXTEND tool from the Trim/Extend flyout on the Modify tool box.

2. Pick the Ground Line. Press the Enter button on the mouse to complete the selection process.

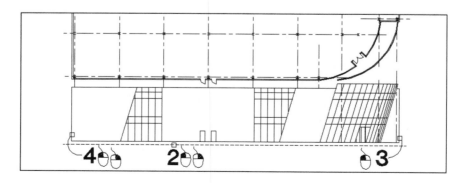

3. Pick the lower half of the right building wall line to extend it to the ground line.

4. Pick the lower half of the left building wall line. Press the mouse's Enter button to end the Extend command.

We now have a schematic elevation and a plan with a strange-looking curved curtain wall floating out in space. To move this object temporarily out of the way, we will put it on a layer we can turn off until we need it again. Start by making a layer named something like Roof_plan. Next pick the Properties tool (DDCHPROP), then pick the curtain wall/roof plan lines with a series of crossing boxes or windows. In the Change Properties dialog box, switch all the objects in the curtain wall to the Roof_plan layer. Now we can make the roof-level curtain wall disappear until we need it again, and we are free to draw some stairs and ramps inside and outside the building.

CHAPTER TWELVE

DESIGNING RAMPS AND STAIRS

Contents

Before we proceed with further plan development, we need to design the exit and lobby stairs, and the exit door ramp on the north side of the building. There will also need to be a ramp on the west end of the building. These designs can then be "plugged into" the plan drawings for the first and second floors. In the next chapter, we will use the elevation we drew in the previous chapter to develop the roof and second-floor plans.

Drawing the Exit Door Ramp

Federal handicap accessibility law and most codes require wheelchair ramps to have a slope not greater than 1 inch for every foot of travel. This relationship means we will need a 24-foot ramp from the first-floor exit door, plus the required landing. The first thing we need to do is move the north elevation down to make room for the ramp.

Move the Elevation to Provide Drawing Room

Use the MOVE tool on the Modify tool box to move the elevation, as shown in the following illustration.

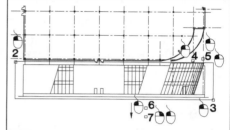

1. Pick the MOVE tool from the Modify tool box.

2. Pick the upper left corner of a Selection Window, making sure it is above the highest part of the building.

3. Pick the lower right corner of the selection window, dragging it completely around the elevation (except for the column grid lines on the right side).

4. Pick one of the column grid lines.

5. Pick the other one. Press the Enter button on the mouse to end the selection process.

6. Pick a point near the elevation. Turn Ortho Mode on.

7. Drag the cursor about 20 feet down (270 degrees), and pick a point. The elevation should now be positioned properly.

Drawing the Plan View of the Ramp

The ramp must first be planned to the length and width required by code and the Americans with Disabilities Act (ADA). Therefore, the first step is to see what these requirements are.

Establish the Ramp Width

The Uniform Building Code requires that a landing for a ramp at an exit door (opening onto it) shall not be reduced to less than 42 inches or by more than 3½ inches from the required minimum width (44 inches) when the door is in the full open position. We will design to the most conservative dimension here: Our ramp width will be 3'-6" (42") plus the 3'-0" extension of the open door (at the leftmost exit), thus giving our ramp a width of 6'-6". An alternate design would be to separate the right exit door with a guardrail and provide it a level walkway to the entry plaza at the front of the building, and narrow the ramp and walkway to 44 inches. Doing that, however, would require that we provide another wheelchair access to the entry plaza at the front of the building, reasonably near the parking area's handicap spaces.

The wider ramp is the least costly solution to both exiting and disabled access to exiting requirements. We will duplicate the ramp in a narrower version to provide access to the building entrance instead and save the cost of the walkway. This allows us to use the same drawing more than once, thus increasing our productivity.

Zoom in on the exit doors as shown in the next illustration. Pull down the Layer List from the Object Properties tool bar and pick Bldgstair as the current layer.

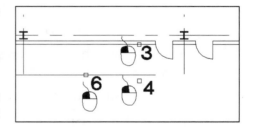

 1. Pick the OFFSET tool on the Modify tool box.

 2. Type 6'2 and press [Enter], either on the keyboard or the mouse.

 3. Pick the outside building wall to the left of the left exit door as the object to Offset.

4. Pick a point below (270 degrees) the wall.

Now we need to change the layer of the line from Bldgwall to Bldgstair:

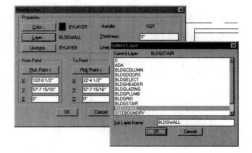

5. Pick the Properties button (DDCH-PROP) on the Main tool bar.

 6. Pick the new line we just Offset.

7. In the Change Properties dialog box, press the Layer button.

8. In the Layer and Linetype Control dialog box, pick the Bldgstair layer. Pick the OK button.

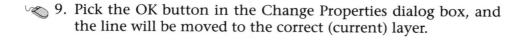
  9. Pick the OK button in the Change Properties dialog box, and the line will be moved to the correct (current) layer.

 1. Pick the LINE tool on the Draw tool box. Pick Tracking on the standard Osnap tool box.

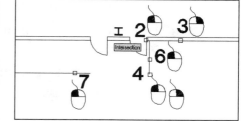

 2. Pick the **Intersection** of the right exit door's right jamb and the wall as the start point for Tracking.

 3. Drag the pointer to the right and Type <u>6</u>, then [Enter]. Press the Enter button on the mouse to end Tracking.

 4. Drag the line down (270 degrees) and pick a point approximately 6'-6" below the building wall. Press the mouse's Enter button to end the Line command.

 5. Pick the FILLET tool on the Modify tool box.

6. Pick the short side of the ramp.

7. Pick the long side of the ramp.

Set the Landing and Ramp Length

 1. Pick the OFFSET tool on the Modify tool box.

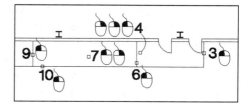

 2. Type <u>15'6</u> and press the [Enter] key. This is the distance required to give us a 5'-0" clear landing beyond the left door as required by the ADA and the UBC.

 3. Pick the right end of the ramp landing.

4. Pick a point to the left of the landing's right end. Press the mouse's Enter button to end the OFFSET command. Press it again to recall the OFFSET command.

 5. Type <u>24'</u> and press the [Enter] key. This OFFSET distance will give us a ramp with the maximum slope of 1 inch per foot allowed by the ADA and the UBC.

6. Pick the left end of the landing.

 7. Pick a point to the left of the landing. We have now established the end of the ramp run.

 8. Pick the FILLET tool on the Modify tool box.

 9. Pick the end of the ramp.

 10. Pick the long side of the ramp.

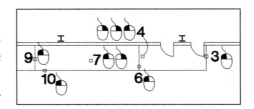

• •

Drawing Handrails in Plans

I generally don't draw handrails and guardrails on plans that are to be plotted at scales smaller than 1/4" to the foot. In design development, however, it is important to deal with issues of appearance and clearances since they are mandated by code and federal law. What we will do here is use single lines to represent the rails in the elevation view and Polylines to show them in plan view so that we can modify the Polylines in the construction drawing details later, and we can cut and paste from the elevation drawing to create the large-scale detail of the handrail/guardrail at the appropriate time. Exterior guardrail design is an issue that must usually be settled with the client early in the design process.

Polylines are used in plan view drawings of stairs and ramps in all my small-scale drawings because it is time wasting to draw two lines 1½" from a wall and then trim countless stair tread lines. With FILL turned on, the polyline accurately represents the width of the handrail and accurately documents design intent. Later chapters on details will deal with the issues of precision and placement of such objects.

Create a new layer called Bldghandrail. If you need to refresh your memory on how to do this, refer to Chapter 3.

Draw the Plan View of the First Handrail

 1. Pick the Polyline (PLINE) tool from the Draw tool box.

 2. Type W to set the Polyline Width, then press [Enter], then type 1.5 and press [Enter] again.

3. Pick the Tracking tool from the standard Osnap tool box or from the customized Draw tool box supplied with this book, on the companion disk. Pick the intersection of the building wall and landing as the first tracking point.

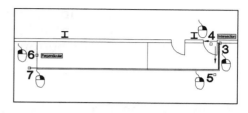

4. Drag the Polyline to the left and type <u>3</u>, and press [Enter]. Press the Enter button on the mouse to end Tracking.

5. Drag the Polyline down (270 degrees) and type <u>5'3</u>; then press the [Enter] key or the mouse's Enter button.

6. Pick the Perpendicular Osnap button from the tool bar. Pick the end of the ramp.

7. Drag the Polyline to the left again and type <u>12</u>, and press the mouse's Enter button.

We have now drawn a plan symbol for a 1 1/2" diameter handrail, inset 3" from the edge of the ramp for graphic clarity when plotted at 1/8" = 1'-0" scale.

Create the Second Handrail in Plan View

The default Polyline Width (PLINEWID) is still 1 1/2" so we won't need to specify it again. Make sure Osnap Mode and Ortho Mode are on.

1. Pick the Polyline tool (PLINE) on the Draw tool box. Pick the Tracking tool from the standard Osnap tool box.

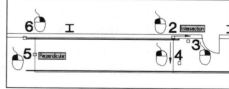

2. Pick the **Intersection** of the landing line and the building wall.

3. Drag the pointer to the right (0 degree) and type <u>12</u> and press [Enter]. Press the Enter button on the mouse to end Tracking.

4. Drag the Pline down (270 degrees) and type <u>3</u>; press [Enter] (key or mouse button).

5. Pick the Perpendicular Osnap button on the Osnap tool bar and then pick the end of the ramp.

6. Drag the Pline to the left (180 degrees) and type <u>12</u>; press [Enter] (key or mouse button).

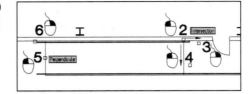

Projecting the Plan of the Ramp to the Elevation

Zoom out to the view shown in the next illustration. The easiest way to accomplish this type of view change is to pick the ⊖ ZOOM Minus button on the tool bar, then use the ⊕ ZOOM Window button to construct the final view.

Make sure Bldgstair is the current layer, and that the ⏏ Running Osnap is set to **Intersection**. Turn the Bldghandrail layer off.

1. Pick the LINE tool from the Draw tool box.

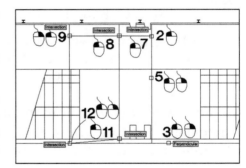

2. Pick the **Intersection** of the lower right corner of the ramp.

3. Pick the Perpendicular Osnap button on the tool bar. Pick the building ground line. Press the mouse's Enter button to end the Line command.

4. Pick the COPY (Duplicate Objects) tool from the Modify tool box.

5. Pick the line you just drew. Press the Enter button on the mouse to close the selection process.

6. Type <u>M</u> followed by [Enter] to select the Multiple Copy option.

7. Pick the **Intersection** of the line and the corner of the ramp for the first displacement point.

8. Pick the **Intersection** of the landing and the ramp.

9. Pick the **Intersection** at the end of the ramp and press the Enter button on the mouse to end the Duplicate Objects (COPY/DUP) command.

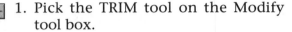 10. Pick the LINE button on the Draw tool box again.

11. Pick the intersection of the middle construction line and the ramp landing in the elevation.

12. Pick the intersection of the construction line at the landing's end and the ground line. Press the mouse's Enter button to end the Line command.

Complete the Ramp Elevation

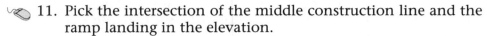

1. Pick the TRIM tool on the Modify tool box.

2. Pick the floor line on the elevation of the north side of the building. Press the Enter button on the mouse to end the selection of cutting objects.

3. Pick the right construction line as the line to be Trimmed. Press the mouse's Enter key to end the Trim command.

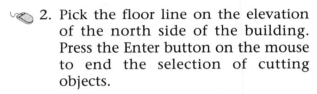 4. Pick the ERASE tool on the Modify tool box, or type <u>E</u> and press [Enter] on the keyboard.

5. Pick the remaining construction lines. Press the mouse's Enter button to end the Erase command, or select the lines and press the [Delete] key.

Drawing the Ramp Handrails and Guardrails in Elevation

In creating the ramp's guardrail, we are going to use a number of drawing objects that are not on the Bldghandrail layer. We will use Release 14's powerful Match Properties button to change them once we have created the top guardrail and handrail.

For the design of the guardrail, I have decided to create a simple welded tube rail system. I could have used any number of exterior quality railing systems offered by the major manufacturers for this example, but they all present the same drawing problems, and everyone will end up drawing this more generic rail type at some time or another. Remember too, that at this point we're just creating a single-line schematic representation of the real thing.

Create the Top Guardrail and Handrail

 1. Pick the OFFSET tool on the Modify tool box.

 2. Type 3'6, then press [Enter] for the Offset Distance.

 3. Pick the Floor line of the north elevation.

 4. Pick a point above the Floor line.

 5. Pick the angled surface of the ramp.

 6. Pick a point above the ramp. Press the mouse's Enter button to end the Offset command. Press the Enter button on the mouse again to recall the OFFSET command.

 7. Type 6, then press [Enter] for the OFFSET Distance.

 8. Pick the guardrail above the ramp.

 9. Pick a point below the guardrail, and press the Enter button on the mouse to end the OFFSET command.

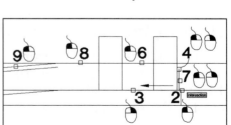

Drawing the Guardrail Posts

Make sure Ortho Mode is on, and that Bldghandrail is the current layer.

 1. Pick the LINE tool from the Draw tool box. Pick the TRACKING tool from the Osnap tool bar.

 2. Pick the **Intersection** of the right end of the ramp landing to begin Tracking.

 3. Drag the pointer to the left, Type 3, and press [Enter]. Press the Enter button on the mouse to end Tracking.

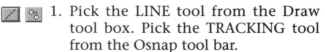

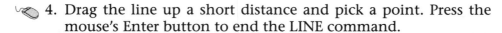 4. Drag the line up a short distance and pick a point. Press the mouse's Enter button to end the LINE command.

5. Pick the FILLET tool from the Modify tool bar.

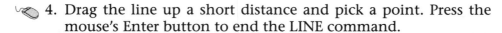 6. Pick the horizontal guardrail line.

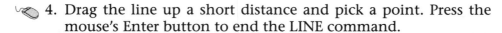 7. Pick the post line we just drew. This will create a 90-degree corner between the post and the handrail. Press the mouse's Enter button to recall the Fillet command.

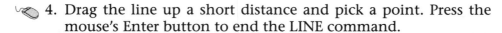 8. Pick the horizontal guardrail line.

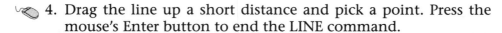 9. Now pick the angled guardrail line to create a joint between the two lines.

The next step is to draw the remaining posts and fillet the handrail and the guardrail at the other end of the ramp. Make sure your Running Osnap is set to **Intersection**.

1. Pick the LINE tool from the Draw tool box.

2. Pick the **Intersection** of the ramp and the landing.

3. Pick the **Intersection** of the horizontal and angled guardrail. Press the mouse's Enter button to end the Line command. Press it again to recall the Line command.

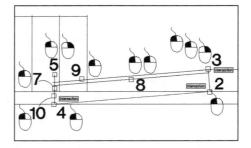

4. Pick the **Intersection** of the ground line and the ramp.

5. With Ortho on, drag the line up (90 degrees) past the guardrail and pick a point. Press the mouse's Enter button to end the Line command.

6. Pick the FILLET tool on the Modify tool box.

7. Pick the post line.

8. Pick the ramp's handrail line. Press the mouse's Enter button to recall the Fillet command.

9. Pick the ramp's guardrail line.

10. Pick the post line again. We now have the guardrail ended with a 90-degree bend.

Draw Handrail Extensions

Now we need to create the code-required 1-foot-minimum extension of the handrail past the end of the ramp and at the landing. Make sure Ortho Mode is still on.

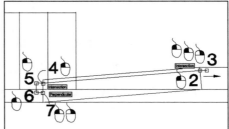

1. Pick the LINE tool from the Draw tool box.

2. Pick the **Intersection** of the right end of the handrail and the post at the top of the ramp.

3. Drag the line to the right and type <u>1'</u> (or <u>12</u>) and press the [Enter] key or the mouse's Enter button. Press the mouse's Enter button again to end the LINE command, and click it once more to restart the Line command.

4. Pick the **Intersection** of the left post and the handrail to start the next rail extension.

5. Drag the line to the left (180 degrees) and type <u>1'</u> (or <u>12</u>) and press the [Enter] key or the mouse's Enter button.

6. Drag the line down (270 degrees) and type <u>1'6</u> (or <u>18</u>) and press the [Enter] key or mouse button again.

7. Pick the Perpendicular Osnap button on the Osnap tool bar and then pick the post. Press the mouse's Enter button to end the Line command.

8. Now pick the FILLET tool on the Modify tool box and FILLET the ends of the handrail and the extensions. Doing this will effectively trim the overlapping ramp handrail line.

Complete the Intermediate Rails and Posts

First we'll make a temporary block and then use the DIVIDE command to locate our posts:

1. Pick the Make Block (BMAKE) tool from the Draw tool box's Block flyout.

2. The Block Definition dialog box will appear. Type the block name of your choice (I used "post"). Pick the Select Object button.

 3. Pick the Select Point button. Pick the **Intersection** of the post and the guardrail. Press the OK button on the dialog box.

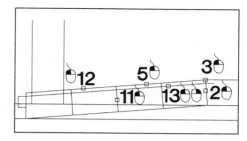

 4. Pick the DIVIDE tool on the Draw tool box.

 5. Pick the guardrail as the object to divide. You must now respond to the following prompts on the command line:

 6. <Number of segments>Block: Type <u>B</u> to select the Block option, and press [Enter].

 7. Block name to insert: Type the name of the block, and press [Enter].

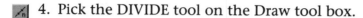 8. Align block with object?<Y>: Type <u>N</u> for No to prevent Auto-CAD from rotating the block.

 9. Number of segments: Type <u>4</u>, and press [Enter]. This will give us three posts almost equally spaced along the ramp, as illustrated.

Using the Match Properties Tool

Because we used the ramp and landing lines to create the top rails, they are not on the Bldghandrail layer. To quickly remedy this situation with three quick mouse picks, we will use a tool new to AutoCAD with Release 14:

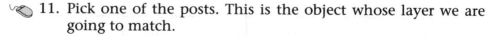 10. Pick the Match Properties tool on the Standard tool bar.

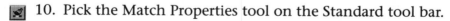 11. Pick one of the posts. This is the object whose layer we are going to match.

12. Pick the guardrail.

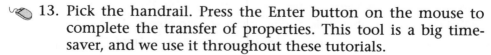 13. Pick the handrail. Press the Enter button on the mouse to complete the transfer of properties. This tool is a big time-saver, and we use it throughout these tutorials.

Now we'll do the intermediate rails:

The UBC requires less than 4 inches of space between guardrail members. I designed the span of the horizontal rails to be short enough to use 1.25-inch-diameter steel tubing, which gives us an on-center dimension of 5 inches as the safe spacing. Zoom in on the handrail, as shown in the next illustration, to make the following object picks easier.

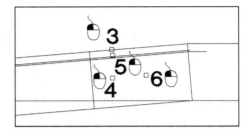

 1. Pick the OFFSET tool from the Modify tool box.

 2. Type <u>5</u> and press [Enter] on the keyboard or mouse.

 3. Pick the guardrail as the object to Offset.

 4. Pick a point below the handrail.

 5. Pick the new rail created by the Offset.

 6. Pick a point below the handrail again.

Continue this process until you have created seven new rails. It's not as tedious to do as it is to read about! Zoom back out so you can see both the complete ramp handrail and some of the landing railing.

Now repeat the OFFSET process for the landing railing. Start by picking the OFFSET tool in the Modify menu again, only this time, instead of typing in the 5-inch dimension, press the mouse's Enter button to accept the current (5") offset.

Pick the top guardrail as the object to OFFSET and continue as before. When you are finished, you should have a result that matches the next illustration.

Note: We could have used the 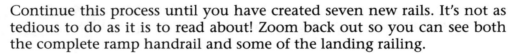 ARRAY tool on the Modify tool box to create eight rows of lines spaced –5" apart, but ARRAY is generally slower than OFFSET when fewer than ten objects are involved in this kind of orthographic duplication. Also, ARRAY of angled (nonorthographic) objects does not space the objects the way OFFSET does. To achieve a true 5" spacing, we would need to change AutoCAD's Snap Angle variable before using the ARRAY command.

Now we're going to clean up the railing, and we will be finished.

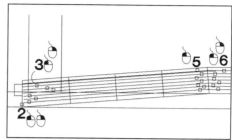

 1. Pick the EXTEND tool on the Modify tool box.

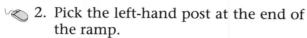 2. Pick the left-hand post at the end of the ramp.

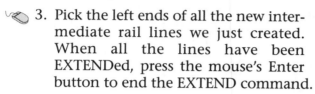 3. Pick the left ends of all the new intermediate rail lines we just created. When all the lines have been EXTENDed, press the mouse's Enter button to end the EXTEND command.

4. Pick the FILLET tool on the Modify tool box.

5. Pick a ramp rail line.

6. Pick a landing rail line. Press the mouse's Enter button to recall the FILLET command. Continue this process until you have FILLETed all seven rail lines.

Designing Stairs

We are first going to design the north exit stair, then migrate that design to the other exits and the lobby. Zoom Window on the north exit doors to create the view shown in the next illustration.

The first thing we need to determine is the distance from the bottom step to the column enclosure (between the exit doors). The column enclosure should clear the steel by at least 4 inches, and perhaps more, because we will need to give this enclosure a 1-hour rating. With the overall size of the fireproofed steel being roughly 17" x 17", a 2'-3"-wide enclosure should do the trick, unless the structural engineer changes the steel size drastically.

Start the North Exit Stair

Turn on the Bldgrid layer. Make sure Ortho Mode is on. Make sure you are drawing on the Bldgstair layer.

 1. Pick the LINE tool on the Draw tool box. Pick the Tracking tool on the Osnap tool bar or the customized Draw tool box.

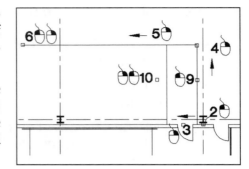

2. Pick the **Intersection** of the column center line and the inside of the building wall as the first Tracking Point.

3. Drag the line to the left and type <u>1'1.5</u> and press either the [Enter] key or the mouse button. Press the Enter mouse button again to end Tracking.

4. Drag the line up (90 degrees), type <u>16'</u>, and then press [Enter].

5. Drag the continuing line to the left (180 degrees) until it is just short of the left side of the screen.

6. Pick the line's endpoint and press the mouse's Enter button to end the LINE command.

Landing Design Considerations

Now we need to design the depth of the bottom landing. We have chosen a 4'-0" overall wide steel drop-in stair with 3/16" steel plate side members. While the stair width may be no less than 44 inches wide by code, and landings need to be only as deep as the stair width, the handrail at the bottom landing will intrude 1'-0" into the landing space plus the depth of one stair tread, or 1'-11". We also need to make sure this required handrail extension does not overlap the door frame.

To check the distance needed, use the ▦ DISTANCE tool to measure the dimension from the vertical line we just drew to the right **Intersection** of the wall and door jamb. Adding the 3'-0" door width to that and 2'-11" plus 2 1/2" for the door frame on the left side tells us that a 6'-0" offset between wall and stair tread will be ample and allow some margin for construction error. The use of a drop in stair helps too because it will be fabricated first and will be installed at the same time steel is erected, with the shaft way framing occurring with all the critical elements in place.

In general, whether you are designing in the United States or not, it is at this stage in the project that regulatory issues such as this must be considered, because they will affect all the drawings that follow. This is why

I advocate inserting the stair drawings in a plan as separate reference drawings, because they are so subject to last-minute change under regulatory review.

 7. Pick the OFFSET tool on the Modify tool box.

 8. Type 6' followed by [Enter]. This is the distance from the column enclosure/stairwell wall to the first riser nosing.

 9. Pick the line we just drew next to the column.

 10. Pick a point to the left of the line. Press the Enter mouse button to end the OFFSET command.

Develop the Schematic Stair Elevation/Section

The first design decision we need to make is the tread depth. In manual drafting, we would study rise and run charts to arrive at this dimension, but AutoCAD will perform most of the calculations for us, allowing us to design using the ADA legal minimum tread depth of 11 inches, so that's what we will select. The first step (pun intended) is to draw an 11-inch tread and make a BLOCK of it.

Drawing the First Tread

 1. Pick the LINE tool from the Draw tool box.

 2. Pick a point in a clear area of the screen. Drag the line in a horizontal direction (0 or 180 degrees).

 3. Type 11 followed by [Enter]. This line represents our standard tread.

 4. Pick the Make Block tool (BMAKE) on the Draw tool box.

 5. Type the block name in the Name space on the Block Definition dialog box (I used the obvious name "tread"). Pick the **Select Objects** button.

 6. Pick the tread line. Pick the **Select Point** button on the dialog box.

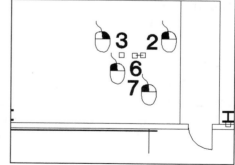

7. Pick the **Endpoint** Osnap button on the Object Snap tool bar. Pick the RIGHT end of the tread line. Make sure the "Retain Objects" box on the Block Definition dialog box is not checked (pick it to remove the checkmark).

8. Pick the OK button on the dialog box.

We must now make sure AutoCAD will operate at the maximum mathematical precision for our next operation. Drop down the Format Menu from the menu bar at the top of the screen. Pick Units on the menu. The Units dialog box will now appear, as shown in the illustration.

1. Drop down the Precision list.
2. Scroll down to the highest value and pick it. Pick the OK button.

When we let AutoCAD calculate the rise of the stair, we want it to use equal height risers, as required by code. If we use a precision as low as 1/16 of an inch, the program will round off its calculation and throw the difference into the top and/or bottom riser, with all the others being equal to one another. Accuracy is what we use a computer for, so we need to make sure we get it when needed.

Using the Computer to Generate the Riser Height

A quick study tells us that 16 feet (our floor-to-floor distance) divides equally into 32 six-inch risers. Six-inch risers are not really desirable, however, so we'll reduce the number to 30 to gain a little more height.

1. Pick the DIVIDE tool from the Modify tool box (it's not on the standard modify tool box; if you haven't added it to your tool box, Appendix A will show you how, or you can select it from the Modify menu).

2. Pick the vertical line to the left of the exit door as the line to divide. Reply to the following command line prompts as shown in the next steps:

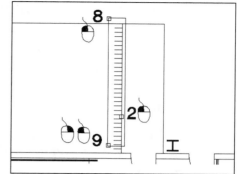

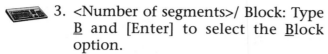 3. \<Number of segments>/ Block: Type <u>B</u> and [Enter] to select the <u>B</u>lock option.

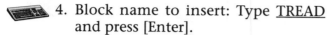 4. Block name to insert: Type <u>TREAD</u> and press [Enter].

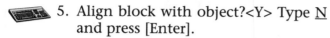 5. Align block with object?\<Y> Type <u>N</u> and press [Enter].

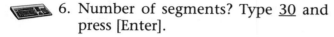 6. Number of segments? Type <u>30</u> and press [Enter].

AutoCAD almost instantly fills the line with TREAD blocks, inserted at equal intervals, as shown in the illustration.

7. Pick the ERASE tool from the Modify tool box.

8. Pick the first point of a selection Window to the upper left of the top tread line.

9. Pick the endpoint for the window to the right of the next to lowest tread line, as shown in the illustration. Press the Enter button on the mouse to complete the erasure. That leaves us with only one tread line remaining, but that's all we need to complete the stair.

Drawing the First Step

Zoom Window on the last tread line until it is as large as shown in the following illustration. We're now designing the step we are going to use, but only schematically. The detail will come much later in the process and may actually be drawn by the stair manufacturer.

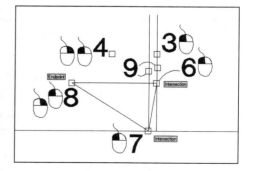

1. Pick the OFFSET tool on the Modify tool box.

2. Type <u>1</u> and [Enter] to set the Offset distance to 1 inch. I generally like to design things in even inches, given the choice. The UBC and the ADA allow a 1 1/2" recess offset of tread to nosing, but 1 inch works fine for me.

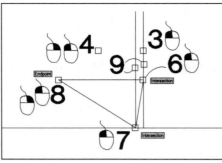

 3. Pick the vertical construction line.

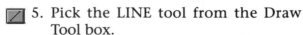 4. Pick a point to the left of the line. We have now created a second construction line.

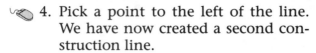 5. Pick the LINE tool from the Draw Tool box.

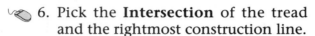 6. Pick the **Intersection** of the tread and the rightmost construction line.

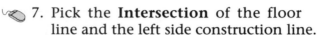 7. Pick the **Intersection** of the floor line and the left side construction line.

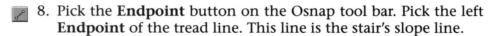 8. Pick the **Endpoint** button on the Osnap tool bar. Pick the left **Endpoint** of the tread line. This line is the stair's slope line.

9. Pick the ERASE tool on the Modify tool box, and erase the two construction lines.

Note: We could have used Tracking and the LINE command to draw the nosing without using a second construction line by picking the LINE Tool, then Tracking, then the **Intersection** of the tread and construction line, then the intersection of the construction line and floor line. At that point, we drag the line to the left and type 1 + [Enter], and press the Enter button on the mouse to end Tracking. Then we have the option of picking the right or left end of the tread to draw either the riser or the slope line. Completing the step would require repeating the LINE command and drawing the remaining riser or slope line. This method has no significant speed advantage over using OFFSET and an additional construction line.

Adding the Run to the Rise

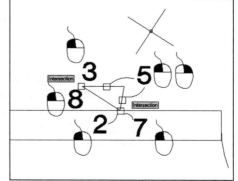

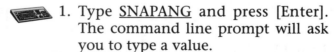 1. Type <u>SNAPANG</u> and press [Enter]. The command line prompt will ask you to type a value.

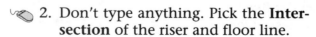 2. Don't type anything. Pick the **Intersection** of the riser and floor line.

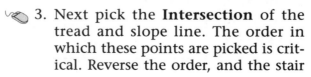 3. Next pick the **Intersection** of the tread and slope line. The order in which these points are picked is critical. Reverse the order, and the stair

will be projected down and to the right, instead of up and to the left, which is what we want.

You will notice that the cursor crosshair has turned to an angle equal to that of the stair slope. The SNAPANG variable controls the angle of the Snap Grid in AutoCAD and is a very handy tool to use if you draw lots of repeated objects at odd angles. It can also be set from the Drawing Aids dialog, which is accessed by typing DDRMODES at the command line, or by selecting it under the Format menu.

4. Pick the ARRAY tool on the Modify tool box. Press the mouse's Enter button to accept the default <u>R</u>ectangular Array.

5. Pick the tread and riser as the objects to ARRAY.

6. Press [Enter] to accept the default value of <1> for the number of rows. Type <u>30</u> and [Enter] when prompted for the number of columns.

7. You will now be prompted for the distance between the columns. Instead of typing, pick the **Intersection** of the riser and floor line.

8. Pick the **Intersection** of the tread and slope line.

AutoCAD will almost instantly draw 30 risers and 29 treads at the correct slope angle.

9. Type <u>SNAPANG</u> and [Enter]. Type <u>0</u> and [Enter] to reset the grid to its' normal orientation.

Zoom out to get a view of the completed stair. Make sure Ortho Mode is on. We are now going to "fold" the stair into three runs to keep the stair shaft area as small as possible.

Fold the Run into the Final Stair Elevation Schematic

1. Pick the MIRROR tool on the Modify tool box.

2. Select the stairs to Mirror by starting a Crossing Box below the eleventh tread from the "floor."

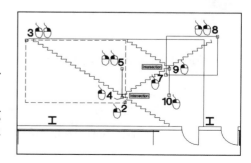

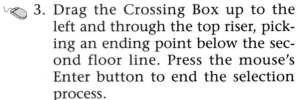

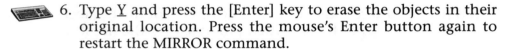

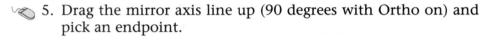

3. Drag the Crossing Box up to the left and through the top riser, picking an ending point below the second floor line. Press the mouse's Enter button to end the selection process.

4. Pick the first point of the mirror axis at the **Intersection** of the tenth tread and the eleventh riser as shown in the illustration.

5. Drag the mirror axis line up (90 degrees with Ortho on) and pick an endpoint.

6. Type <u>Y</u> and press the [Enter] key to erase the objects in their original location. Press the mouse's Enter button again to restart the MIRROR command.

You should now have the upper two-thirds of the stair running in the opposite direction from the bottom third.

7. Select the next section to Mirror by starting a Window just below the tenth tread from the bottom of the second run, as illustrated.

8. Pick the end point for the Window above the top riser. Press the Enter button on the mouse to complete the selection.

9. Pick the start point of the Mirror axis by selecting the **Intersection** of the tenth tread and the eleventh riser.

10. Drag the Mirror axis line down (270 degrees) or up (90 degrees) into open space and pick its endpoint.

11. Type <u>Y</u> and press the [Enter] key to erase the objects in their original location.

Creating the Landings in Elevation

1. Zoom Window on the top of the first run, as shown in the illustration.

2. Pick the top tread of the lower run to activate its Grips.

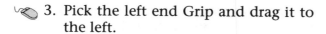

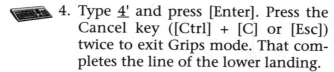 3. Pick the left end Grip and drag it to the left.

4. Type 4' and press [Enter]. Press the Cancel key ([Ctrl] + [C] or [Esc]) twice to exit Grips mode. That completes the line of the lower landing.

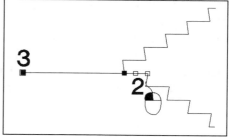

1. ZOOM Previous, or ZOOM in on the stair to get the view shown in this illustration. Pick the MOVE tool on the Modify tool box.

2. Pick the starting point for a selection Window just below the first landing.

3. Pick the endpoint for the selection Window just beyond the top tread of the second run. Press the mouse's Enter button to complete the selection.

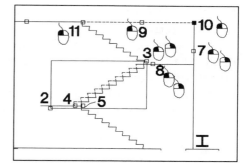

4. Pick the intersection of the first riser of the second run and the landing as the point to move from.

5. Pick the intersection of the first run's top riser and landing as the point to move to.

6. Pick the EXTEND tool from the Modify tool box.

7. Pick the vertical line on the right as the target line. Press the Enter button on the mouse to complete selecting.

8. Pick the top tread of the second stair run. Press the mouse's Enter button to complete the EXTEND command.

9. Pick the second-floor line to turn on its' Grips.

10. Pick the right end grip to make it active.

11. Drag the Grip to the **Intersection** of the top riser and the floor line, and pick the **Intersection**.

There we have it: one schematic stair design elevation in nine steps. All the critical vertical dimensions are set, and the run is nailed down. This is the equivalent of the large-scale sketch we would draw on paper at this stage in the design development process, but unlike that paper sketch, our stair will *not get redrawn* in a final detail section. The schematic CAD drawing will become the final detail section as the drawings develop. Next we will use the elevation to generate the first- and second-floor plan drawings of the stair and shaftway.

Draw Final Plan Views of the Stair

It's now time to determine what type of drop in stair we plan to use. I have opted for a welded steel plate design using 3/16"-thick plate stringers and a minimum 14-gauge formed steel pans and risers. I've also decided to keep the stair shaft interior walls 4" away from the stair, to allow for easier finishing of the drywall. The stair runs will be separated by 8 1/2" between landings. The 2-hour shaftway wall is a nominal 5 1/2" thick, 5 5/8" actual thickness.

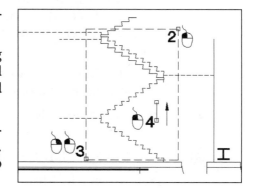

1. Pick the MOVE tool from the Modify tool box.

2. Pick the starting point for a Crossing Box above the second floor line and beyond the top riser of the second stair run.

3. Pick the second point of the Crossing Box as shown in the illustration. Press the mouse's Enter button to complete the selection.

4. Pick a first displacement point in an empty part of the screen.

5. Drag the pointer up 90 degrees (with Ortho Mode on). Type 9'-10 and press [Enter]. This moves the stair elevation to the outside surface of the shaftway wall.

Draw the Plan of the Stair Shaft

Make sure Ortho Mode is on, and that the current layer is Bldgwall.

1. Pick the LINE tool from the Draw tool box.

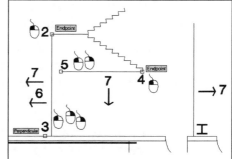

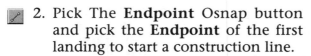 2. Pick The **Endpoint** Osnap button and pick the **Endpoint** of the first landing to start a construction line.

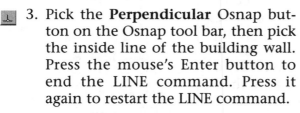 3. Pick the **Perpendicular** Osnap button on the Osnap tool bar, then pick the inside line of the building wall. Press the mouse's Enter button to end the LINE command. Press it again to restart the LINE command.

4. Pick the **Endpoint** of the first riser in the stair elevation.

5. Drag the line to the left, as illustrated, and pick an endpoint for it. Press the mouse's Enter button to end the LINE command.

6. Pick the OFFSET tool on the Modify tool box. Type <u>4</u>, followed by [Enter]. This is the distance from the landing to the shaft wall. Pick the vertical line we just drew, and OFFSET it to the left. Press the mouse's Enter button to end the OFFSET command, then press it again to restart OFFSET.

7. Type <u>5.5</u> and press [Enter]. This is our nominal shaft wall width. OFFSET the two vertical lines and one horizontal line in the directions shown in the illustration. Press the Enter button on the mouse when you are finished.

8. Pick the FILLET tool on the Modify tool box. FILLET the inside lines of the shaft wall first, moving in a clockwise or counter-clockwise direction (be consistent and filleting will be most efficient). Continue the process for the outside walls. Press the mouse's Enter button to end the FILLET command.

Your stair shaft walls should now look like those in the next illustration.

Create the First-Floor Stair Plan

Make Bldgstair the current layer. Make sure Ortho Mode is on and that Running Osnap is set to **Intersection**.

1. Pick the LINE tool from the Draw tool box.

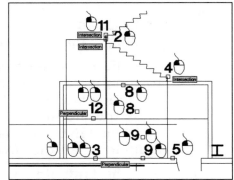

2. Pick the **Intersection** of the first landing and the riser below it.

3. Pick the **Perpendicular** Osnap button on the Object Snap tool bar. Pick the inside line of the building wall, at the bottom of the screen. Press the mouse's Enter button to complete the LINE command, then press it again to restart the LINE command.

4. Pick the **Intersection** of the lowest tread and riser to start the next line.

5. Pick the **Perpendicular** Osnap button on the Object Snap tool bar. Pick the inside line of the building wall, at the bottom of the screen. Press the mouse's Enter button to complete the LINE command.

6. Pick the OFFSET tool on the Modify tool box.

7. Type 4'4 and press [Enter]. This is the distance from the shaft wall to the inside of the stair.

8. Pick the inside stair shaft wall as shown in the illustration. Pick a point below it to OFFSET the wall line in the 270-degree direction.

9. Pick the inside building wall, then pick a point above it to OFFSET the line in that direction (90 degrees). Press the Enter button on the mouse to end the OFFSET command.

10. Pick the LINE tool on the Draw tool box again.

11. Pick the nosing **intersection** above the landing.

12. Pick the **Perpendicular** Osnap button. Drag the line down (270 degrees) and pick the inside stringer line for the upper stair section.

Now we will trim the mess we just created:

1. Pick the TRIM tool on the Modify tool box.

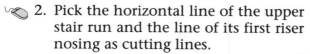 2. Pick the horizontal line of the upper stair run and the line of its first riser nosing as cutting lines.

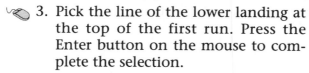 3. Pick the line of the lower landing at the top of the first run. Press the Enter button on the mouse to complete the selection.

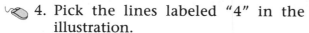 4. Pick the lines labeled "4" in the illustration.

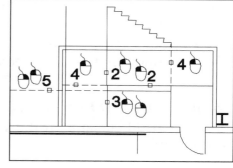

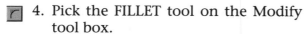 5. Pick the inside line of the lower stair run, outside the shaftway, as illustrated. Press the mouse's Enter button to end the TRIM command.

Completing the Stair Plan Outline

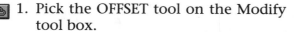 1. Pick the OFFSET tool on the Modify tool box.

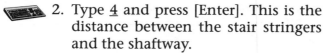 2. Type <u>4</u> and press [Enter]. This is the distance between the stair stringers and the shaftway.

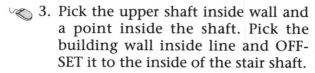 3. Pick the upper shaft inside wall and a point inside the shaft. Pick the building wall inside line and OFFSET it to the inside of the stair shaft.

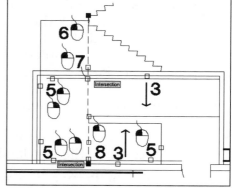

4. Pick the FILLET tool on the Modify tool box.

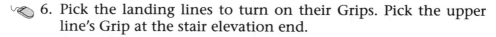 5. FILLET the outside stair lines, starting at one end and moving in a consistent clockwise or counterclockwise direction. Pick the trailing line first, then the line ahead. Once the FILLET is done, press the mouse's Enter button to restart FILLET and pick the last line previously picked again. Move to the next intersecting line and pick it. You will be surprised at how fast this process is, compared with using the TRIM command.

6. Pick the landing lines to turn on their Grips. Pick the upper line's Grip at the stair elevation end.

7. Drag it down to the intersection of the outside stair stringer and landing line.

8. Repeat with the other landing line, dragging its lower endpoint up to its intersection with the outside stringer.

Completing the First-Floor Stair Plan

Draw the tread cut line on the second run (Ortho Mode can be off for this operation).

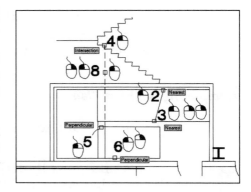

1. Pick the LINE tool on the Draw tool box.

2. Pick the NEARest Osnap button on the Object Snap tool bar. Pick a point on the upper stair run's stringer.

3. Pick the NEARest Osnap button again and pick the inside stringer as illustrated. Press the mouse's Enter button to complete the LINE command. Press it again to restart LINE.

Draw the next two tread lines.

4. Pick the **Intersection** of the tread and riser in the elevation just below the first landing.

5. Pick the **Perpendicular** Osnap button on the Object Snap tool bar. Pick the inside stringer line of the lower stair run.

6. Pick the **Perpendicular** Osnap button again. Pick the outside stringer line. Press the mouse's Enter button to end the LINE command.

7. Pick the ERASE tool on the Modify tool box.

8. Pick the new line we just drew, as shown in the illustration. Press the Enter button on the mouse to complete the erasure.

I wish Tracking worked with Osnap commands, but at this writing it doesn't. Actually, the drawing of the two line segments is almost as fast as using Tracking, and it's true that you can visualize what you are doing a little better. Next we are going to ARRAY and TRIM the tread lines to complete the stair plan.

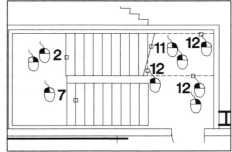

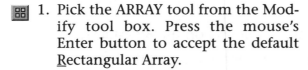 1. Pick the ARRAY tool from the Modify tool box. Press the mouse's Enter button to accept the default <u>R</u>ectangular Array.

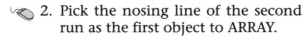 2. Pick the nosing line of the second run as the first object to ARRAY.

3. Press the mouse's Enter button to accept the default <1> value for the number of rows.

4. Type <u>10</u> for the number of columns and press [Enter].

5. Type <u>10</u> (only 10" of the 11" tread is visible) for the distance between columns and press [Enter]. Press [Enter] again to recall the ARRAY command.

6. Press the Enter button on the mouse to accept the default <u>R</u>ectangular Array.

7. Pick the first tread line (not the landing line) nearest the upper landing of the first run as the object to Array.

8. Type <u>9</u> for the number of columns and press [Enter].

9. Type <u>10</u> for the distance between columns and press [Enter].

10. Pick the TRIM tool on the Modify tool box.

11. Pick the angled cut line we drew in the previous step and press the mouse's Enter button to complete the cutting edge selection.

12. Pick the extensions of the stringer lines beyond the cut line, and pick the tenth stair tread/landing line's lower end, as illustrated. Press the mouse's Enter button to end the TRIM command.

There it is, a completed ground-floor stair plan. This drawing will become the basis for our upper-floor stair plan, but first we need to untangle the messy layer situation we have created, and then save the stair as a block definition in the drawing.

Moving Everything to the Correct Layer

 1. Pick the Match Properties button on the Object Properties tool bar.

 2. Pick a stair tread line that is obviously on the Bldgstair layer.

 3. Pick all the other stair components that are not on the Bldgstair layer, and then press the mouse's Enter button. Press the Enter button again to recall the Match Properties command and repeat the process for any shaftway wall lines that are not on the Bldgwall layer.

Create the First-Floor Stair Plan Block

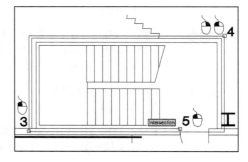

 1. Pick the Make Block (BMAKE) tool from the Draw tool box.

 2. Type a descriptive block name such as stair-plan1, and pick the Select Objects button.

 3. Pick the starting point of a selection Window below the lower left corner of the stair shaft wall, but inside the building wall.

 4. Pick the endpoint for the Window just outside the upper right corner of the stair shaft. Press the mouse's Enter button to confirm the selection.

 5. Pick the Select Point button on the Block Definition dialog box. Make sure the Retain Entities box is *checked*. Pick the **Intersection** of the inside left corner of the exit door jamb for the block insertion point.

 6. Press the OK button on the Block Definition dialog box, and the block is created inside the drawing.

It's not necessary to write the block out to a separate file at this time, but pick the 🖫 SAVE button on the Standard tool bar to save the drawing, thus preserving the block definition in the current file, just in case.

Draw the Second-Floor Stair Plan

This is one instance where CAD drafting is more productive than manual drafting. In two steps, we're going to transform one plan drawing into another, a process that would require line-by-line tracing and redrawing on paper.

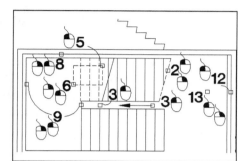

1. Pick the MOVE tool on the Modify tool box. Make sure Ortho Mode is on.

2. Pick the diagonal cut line as the object to move. Press the Enter button on the mouse to end the selection.

3. Pick a point below the line, between the stairs. Drag the line into the position shown in the illustration.

4. Pick the ERASE tool on the Modify tool box.

5. Pick the starting point for a Crossing Box to the right of the first three stair lines as shown in the illustration.

6. Drag the Crossing Box down and to the left until it passes over the three stair lines. Pick its lower left endpoint. Press the mouse's Enter button to complete the erasure.

7. Pick the EXTEND tool on the Modify tool box.

8. Pick the upper inside stair shaft wall line as the object to Extend TO.

9. Pick the lower landing line on the right side, and pick the left side landing line on its upper end. Press the mouse's Enter button to end the EXTEND command.

10. Pick the OFFSET tool on the Modify tool box.

11. Type 4 and press [Enter].

12. Pick the right inside stair shaft wall line, as illustrated.

13. Pick a point to the left of the wall. Press the mouse's Enter button to end the OFFSET command.

Now for the last step:

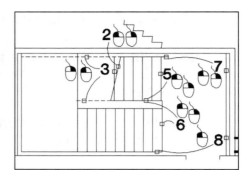

 1. Pick the TRIM tool on the Modify tool box.

 2. Pick the diagonal cut line as the trimming edge. Press the mouse's Enter button to complete the edge selection.

 3. Pick the stringer lines of the lower stair run as shown in the illustration. Press the mouse's Enter button to end the TRIM command.

 4. Pick the FILLET tool on the Modify tool box.

 5. Pick the partial tread line on the lower landing, then pick the lower side stringer for the same stair. Press the Enter button on the mouse to recall the FILLET command.

 6. Pick the same stringer line again. Now pick the line of the lower landing, as illustrated. Press the Enter button on the mouse to recall the FILLET command.

 7. Pick the upper stringer line of the lower run and the right landing line we just created with the OFFSET command. Press the Enter button on the mouse to recall the FILLET command.

 8. Pick the same landing line again, and pick the lower stringer line for the top run.

Adding the Second-Floor Exit Door

That's all there is to it. Now we need a door to the stair on the second-floor level. If you skipped Chapter 9 on drawing doors, read it now to draw this one. Hint: It's easiest to use OFFSET to create the opening jamb lines by Offsetting the left inside shaft wall line 6 inches and then Offsetting the new left jamb line 3 feet. Grab wall and jamb lines with a Crossing Box using the TRIM command to clip the jambs and wall.

Cleaning Up the Layers Again Using Match Properties

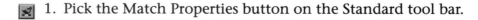 1. Pick the Match Properties button on the Standard tool bar.

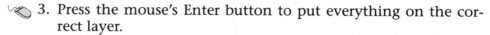

2. Pick an object on the Bldgstair layer. Pick all the objects that should be on that layer but aren't, such as the rightmost landing line of the lower landing.

3. Press the mouse's Enter button to put everything on the correct layer.

Repeat the above work for anything else that needs to be on another layer.

Create a Second-Floor Stair Block

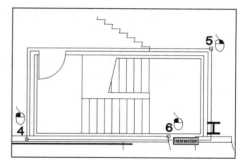

1. Pick the Make Block (BMAKE) tool from the Draw tool box.

2. Type a descriptive block name such as <u>stair-plan2</u>, and pick the Select Objects button.

3. Pick the starting point of a selection Window below the lower left corner of the stair shaft wall, but inside the building wall.

4. Pick the endpoint for the Window just outside the upper right corner of the stair shaft. Press the mouse's Enter button to confirm the selection.

5. Pick the Select Point button on the Block Definition dialog box. Make sure the Retain Entities box is NOT checked. Pick the **Intersection** of the inside left corner of the exit door jamb for the block insertion point.

6. Press the OK button on the Block Definition dialog box, and the block will be created in the drawing database but erased on the screen.

7. Pick the Save Drawing (SAVE) button on the Standard tool bar to SAVE THE DRAWING NOW.

Insert the First-Floor Exit Stair in the Plan

The last step is to insert the stair-plan1 block back in the drawing at all the exits. Make sure that your Running Osnap is set to **Intersection**, and that Ortho Mode is on.

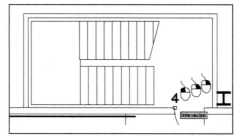

 1. Pick the Insert Block (DDINSRT) tool on the Draw tool box.

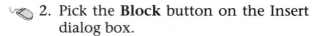 2. Pick the **Block** button on the Insert dialog box.

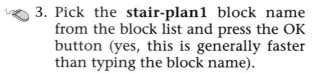 3. Pick the **stair-plan1** block name from the block list and press the OK button (yes, this is generally faster than typing the block name).

4. Pick the intersection of the inside building wall and the left side of the stair's exit door jamb as the insertion point. Press the mouse's Enter button twice, and the stair will be correctly positioned.

This is just the start of working with stairs. Chlapter 13 will show how to build on what we have just drawn to create other stairs in the entrance lobby, and the entrance steps and ramp at the front of the building.

CHAPTER THIRTEEN

FINAL DESIGN DEVELOPMENT OF STAIRS AND BUILDING ELEVATIONS

Contents

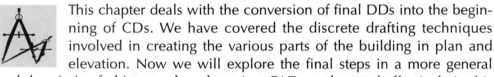

This chapter deals with the conversion of final DDs into the beginning of CDs. We have covered the discrete drafting techniques involved in creating the various parts of the building in plan and elevation. Now we will explore the final steps in a more general and descriptive fashion, to show how AutoCAD can be used effectively in this phase of the work.

By now it has probably occurred to you more than once that we have been investing a considerable amount of time in precision drawing where we would normally be sketching a good deal of the design freehand, or at least with no more than rough approximation of the final CD (construction drawing) precision, if we were drawing by hand. This is the greatest change in discipline that CAD drawing imposes on the design development process. It's important to draw precisely on CAD from the beginning because any DD (design development) work that sticks to the screen never has to be revised to create a final CD. Changes to precisely drawn DD objects are easier because you are actually changing a "real" object to something equally precise.

Manual DDs need to be completely redrawn as precisely calculated CDs just to confirm design intent. When problems crop up with making the intent precise, redesign is often the only solution. CAD drawings, however, can often bypass the "now we'll make it real" stage from the beginning. Drawing precisely to general or conceptual dimensions, as we have tried to demonstrate in the previous chapters, now allows us to make simple final adjustments from one known object assembly to another without redrawing and recalculating large parts of the building.

A good example of this is our 8-inch-thick building wall. Suppose we had originally intended to clad most of the exterior with precast aggregate panels on steel framing. Now budget restraints force us to redesign for Dryvit exterior facing. We need to reduce the wall thickness by 1½ inches and move the ramps along the building wall in by the same dimension. On a manual drawing, we would not even consider erasing the perimeter skin for such a minor revision; we would just change the overall outside dimension on the drawing to 3 inches less than before. We can do the same with a CAD drawing, but I advocate putting the "faked" dimension on a separate layer and color so it is easily identified and tracked.

How to "Fake" a Dimension

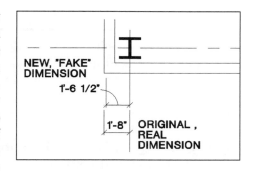

In the following illustration, you can see how to use "faked" dimensions to document a change in the drawing without drawing the change. The original distance from column center line to outside building wall was 1'-8". If our building wall decreases in thickness by 1½ inches, we simply pick the dimension to turn on its Grips, and with Osnap turned off and Ortho Mode on, we drag the origin grip for the left extension line to the right, type 1.5, and press the mouse's Enter button.

As you can see, the dimension now reads 1'-6 1/2". This method is acceptable where few other drawing objects or major geometries are affected. As a general practice, however, it is almost always better to change the *drawing*, especially if other people are going to be using it without reference to your dimensions.

Later, if we use the CAD drawing to calculate the Gross Building Area to BOMA (Building Owners and Managers Association) standards, we will know to make our polyline area boundary of the building 1½ inches smaller in all directions than the walls in the drawing. The CAD system also allows us to MOVE the exterior wall line and anything associated with it 1½ inches with relative ease if the building design is simple and fairly uniform from floor to floor. The important thing is that we have a precise thing to change *from* and another precise thing to change *to*.

Designing the Building Lobby Stair

By using the blocks we have already created for the exit stairs, we can create the more complicated lobby stair in five simple steps (pun intended).

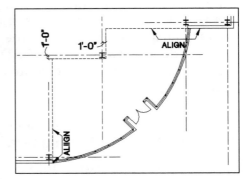

■ ■ First we change the current layer to Intsoffit and turn the Bldgcolumn layer off. Then we draw the line for the second-floor atrium balcony, starting by drawing the lines that align with the ends of the building walls. Next we draw the remaining lines using Tracking from the column grid line Intersection one foot out from the Intersection in both directions.

Note: An alternate method is to use ■ OFFSET to create copies of the grid lines 1'-0" away, then use the Match Properties tool to quickly change their layer and linetype.

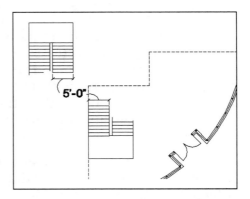

Finally, we ■ FILLET all the corners and we are ready to insert the stairs. To help with this, turn off the Bldgrid and Bldgcolumn layers.

■ Insert both the stair-plan1 and stair-plan2 blocks, rotating and positioning them as shown in the illustration. ■ EXPLODE the blocks.

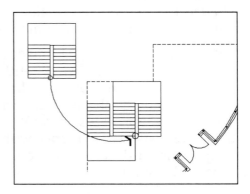

Now ■ MOVE the outside stringer lines 1'-0" to widen the stairs to 5'-0". ■ EXTEND the riser lines using Fences to select the groups of lines. ■ ERASE the cut lines.

■ ERASE the lower risers and cut line from stair-plan2, and use ■ FILLET to complete the upper set of stairs in stair-plan1.

✛ MOVE the stair-plan1 group into position as shown in the illustration.

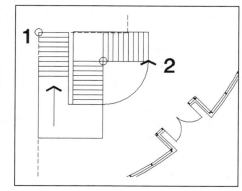

> ✛ 1. MOVE the top run of steps to the line of the second floor-atrium balcony as shown. Then ⊣ EXTEND the landing line to the lowest riser.
>
> ↻ 2. ROTATE the lower run of stairs 90 degrees as shown.

This makes a nice pedestrian (yes, another intended pun) stair design, but our design director is not satisfied with things orthogonal, so we will use one more operation to kink the stair a bit.

↻ Here we use the ROTATE tool one more time to change the angle of the lower landing and lower the runs of stairs to 348 degrees, or 12 degrees from the top run. We then use Grips to shorten the line of the top landing, and we ⌐ FILLET the outside lines.

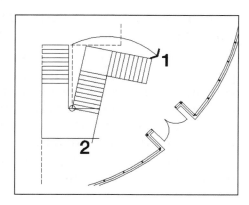

You can use the geometry engine in AutoCAD to play with angles all day, and then in a few quick operations, easily dimension the most radical assembly of forms. Here is where CAD truly blurs the line between design and drafting, if the *designer* is careful to complete the geometric objects and not leave lines scattered about the drawing.

Creating Building Design Sections

An example of section drawing is shown here to give you an idea of generally how to turn an exterior elevation into the basis for a quick section view. We are going to draw a section of the building lobby/atrium, looking north (toward the bottom of the plan).

🔺 The first step is to reverse the plan and elevation, using the MIRROR tool. Make sure Ortho Mode is on before you start. We have drawn midpoint axis lines on the plan, as you can see in the illustration, and have used them as our vertical MIRROR axis. This gets the elevation oriented properly, as if we were inside the building looking out to the north (well, almost).

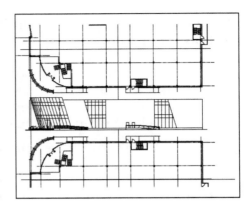

The next step is to mirror the plan again, this time horizontally, using an axis drawn horizontally through the elevation, about 10 feet above the elevation's floor line. This operation produces the result shown in the illustration.

Remember, you can easily UNDO a mirror operation that doesn't work and try again.

We now have the plan set up for our section's point of view. To help with the visualization and projection process, we will draw a section line on the plan and ERASE the upper version of the plan. The next step is to erase those portions of the curtain wall that are outside the section line. By projecting LINES up from the plan, we can draw the doors and the second-floor balcony.

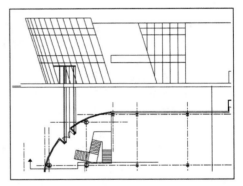

Inserting the Stair into the Section

To make this task easier, we have turned off the layers Bldglazing and Bldgrid, and we have Frozen Bldgcolumn (because the column symbols are blocks).

LINEs are drawn from the stair plan's cardinal points, up to the section. Next the Stair_elev block is inserted on the section's floor line, then MOVEd so that the first tread intersects the line drawn from the first tread on the plan. The Stair_elev block is MIRRORed to the position illustrated, using the vertical construction line at the first tread as the axis, and lastly the block is EXPLODEd.

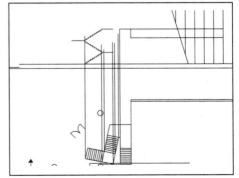

The second landing LINE is drawn from the Endpoint of the stair elevation's landing using Tracking, Perpendicular to the construction line projected from the left corner and then, with Tracking off, perpendicular to the next projected line.

Drawing the Section View of the Stair

LINEs are projected from the end of each of the lower risers, then each riser and tread is MOVEd Perpendicular to these vertical lines. The FILLET tool is used to clean up overlapping lines after the projected construction lines are erased. This is done to project the foreshortening of treads in the section view. We do not change the riser angle due to the imperceptible difference in graphic appearance.

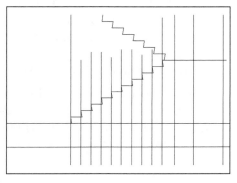

The middle run of stairs has its risers projected to the line projected from the top right riser. Run angle LINEs are drawn from the top riser (landing) line to the top of the lowest riser. A single vertical riser LINE is drawn, then COPIED Multiple times to each intersection of the run angle line and tread. Vertical lines are used because at this narrow angle, the risers' face angle will be so minimal that the eye would see a vertical line anyway.

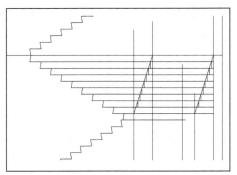

Lastly, the riser lines are TRIMMed and the remains of the Stair_elev block are erased. Now we can turn on the Bldglazing and Bldgrid layers, and thaw Bldgcolumn.

The next step is to project the columns up to the section by drawing LINEs using Tracking to draw them Perpendicular to the floor, then ending Tracking and completing the line in the section. With a considerable bit of TRIMming later, our section drawing will begin to take shape.

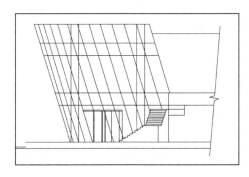

Here's the completed section: We have added the guardrail at the second floor, given the roof of the clerestory a 2:20 slope, added drywall clad beams between the columns, and handrails and guardrails to the stairs, and done much more. Dimension was added to the mullions of the curtain wall by OFFSETting them first 2, then 4 inches. Once they were TRIMmed at the horizontal frames, the mullions closer to the right had their left sides MOVEd to the right to correspond with the amount of surface actually visible.

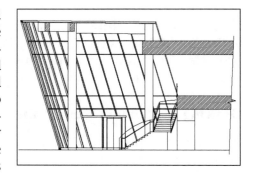

There's still more that could be done, such as showing sections through floor slabs and structural members in the hatched areas, and adding handles to the doors. This drawing, however, is good enough to use in the construction drawing set for cutting details and indicating finishes.

Designing the Exterior Stairs and Handrails

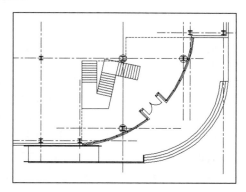

I'll admit it, I took the easy way out in designing the exterior entrance steps and handrails. I simply copied the exit ramp on the north wall, and dragged it into the position you see in the plan. The stairs were constructed with Ortho Mode on, by drawing construction lines from the upper building wall and the corner of the ramp. I then OFFSET the horizontal line a distance equal to the length of the line from the corner of the ramp's landing Perpendicular to the vertical construction line.

Then I drew a symmetrical ARC from the end of the upper wall perpendicular to the lower edge of the ramp landing, using the ARC Start, End, Direction tool. Creating the steps was a matter of OFFSETting the arc 11 inches.

Drawing the Stairs in Elevation

A single horizontal LINE was drawn on the Bldgstair layer from the base of the ramp landing perpendicular to the end of the rightmost building wall. It was ⬛ OFFSET 6 inches repeatedly to form the steps and entrance plaza. The right-most building wall was ⬛ OFFSET 11 inches three times, and changed to the Bldgstair layer using the ⬛ Match Properties tool. All the lines were then ⬛ TRIMmed to form the steps.

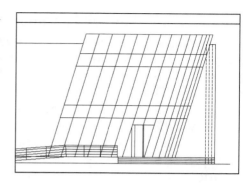

Drawing the Handrails in Plan

The handrails were almost as easy. I created construction lines to locate the ends of the ADA-required 12-inch extensions by OFFSET-ting the upper and lower riser arcs by a foot. Then the prototype handrail was drawn from the Midpoint of the inner construction arc to the Midpoint of the outer one. The construction arcs were erased, and the rail line was turned into a Block with its insertion point at its Intersection with the outermost riser arc. Once the block (named handrail) was made, it was just a matter of using the ⬛ MEASURE tool to Insert the Blocks 8 feet apart (as required by the UBC) by selecting the outer-most riser as the object to MEASURE.

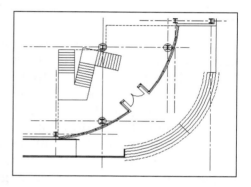

Drawing the First Handrail Elevation

⬛ The first handrail elevation was created by projecting construction LINEs (drawing LINEs using the LINE command, not the Construction Line tool) down from the plan to position the elevation at the top of the arc, parallel to the drawing's *x* axis. A LINE was drawn from the base of the lower step to the base of the top riser, then MOVEd 3'-0" at 90 degrees (Ortho on), and EXTENDed to the inside construction lines. Horizontal LINEs were drawn

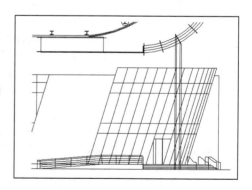

from the ends of the angled handrail line to the outer construction lines, the ⬛ FILLET radius was set to 3 inches, and then the FILLET tool was used to trim the outer construction lines so they became the vertical posts. The last step was to ERASE the inside construction lines.

Shrinking the Handrail Elevations to Simulate Rotated Views

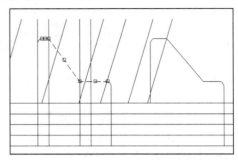

With the first handrail drawn in orthographic elevation, we can now "compress" copies of it into the rest of the rotated handrails. The first step is to create Construction lines to give us a compression boundary to shove the handrail drawing into. You can use the Construction Line tool to do this, snapping the lines to the Intersections and Endpoints of each handrail in the plan. We have illustrated just one set of construction lines here for the sake of clarity.

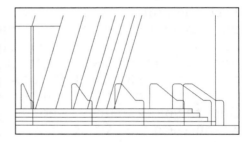

Now we COPY the last drawn handrail, using its Intersection with either the top or bottom landing line as the first point and the Intersection of the corresponding Construction line and landing as the point to COPY to. Next we pick the entire handrail and post assembly with a Window, to turn on its Grips. With Ortho Mode on and Running Osnap set on Perpendicular, we can now easily grab the Grip points of the post and handrail outside the construction lines and move them into place.

The result is a passable representation of the "rotated" object. Note that all of this was accomplished in a few minutes in 2D, without the creation of any Paper Space viewports, 3D models, or other features of AutoCAD. We cheated, of course, because we did not change the radius of the Filleted ends of the rails and posts. When the elevation drawing is plotted in 1/8" = 1'-0" scale, this detail will make no difference.

If we are going to use this drawing for a larger scale plot, or for presentation purposes, then we can create a new file from it and add detail—for instance, we can represent the rails with double lines. The elevation of the handrail also provides us with the basic geometry of the rail that can become the start of our final construction details.

We have essentially completed the north elevation. Two tasks remain, however:

First, we need to create a BLOCK of the elevation drawing (BMAKE / BLOCK) and write it out to a disk file using the WBLOCK command.

I named my file North_Elev.dwg, which is what it is called on the companion disk. I made the insert point the intersection of the ground line and the leftmost wall line.

Second, we need to transfer the window design on the north wall to the plan. 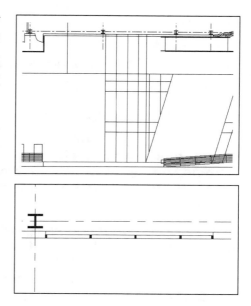 This is done by EXTENDing the window mullion and vertical side lines to the inside building wall and drawing a ▨ LINE from the floor intersection of the angled side.

Then we draw a glass line and ▨ INSERT the Mullion block we created earlier on a construction line OFFSET 2 inches inside the glass line. Once the block is inserted, use Multiple ▨ COPY to position the rest of the mullions.

Repeat this process for the remaining window wall, and the elevation drawing process will be complete for the north side of the building. Notice that we have used the elevation design process to work back and forth with the plan, using one to generate the other.

The Major Difference between Drawing Smart and Drawing with an Architectural Front End Program

Architectural Front End programs (AFEs) come in many flavors and with many levels of features. The least desirable of the lot try to make design decisions for you, by forcing you to pick from built-in libraries of pre-drawn parts, such as windows. As you develop the building plan, you pick windows from the library and insert them in the walls. If you need a window design not in the library, you have to draw the window from scratch, both in elevation and plan, before it can be used. This results in a lot of switching between different views, making changes, inserting windows from the library, and drawing new substitutes.

The supposed efficiencies conferred by such software tend to be illusory if you are doing any type of original design, because of the constraints imposed by the way the software works. I'm not saying AFEs don't have their advantages, but they can generally get in the way more than they help.

The key to efficient use of AutoCAD is to draw nothing before it is needed, and to make *design decisions along with drawing decisions*. You should be aware by now that we have *drawn* everything only once, then made copies and Blocks, altering the original to make it look different.

Completing the Elevations and Assembling the Elevation Drawing

Drawing the West, East, and South Elevations

These elevations are somewhat more straight-forward than the north elevation because they have no sloping walls. All are developed the same way, as illustrated here for the west elevation.

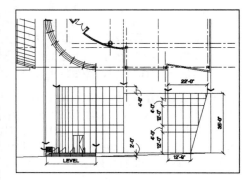

First, create a roof line, then ⬚ OFFSET it 35'-0" to form the floor line. Continue to OFFSET horizontal lines until all the major horizontal planes of the building (and stairs) are established, then project the vertical planes down to the floor line. ⬚ TRIM at this point just to keep things visually organized.

Repeat this process for the detail such as the mullions in the curved atrium curtain wall and the handrails. ⬚ COPY the handrails from the north elevation, ⬚ ROTATE and ⬚ MIRROR them, and ⬚ MOVE them into position.

⬚ Now turn on the Siteboundary layer and project lines (or Construction Lines) down from the site boundary line. Draw a LINE from the building's north corner, at the base of the stairs, Perpendicular to the north boundary projection. Continue by dragging the line down (270 degrees) and typing 3' and pressing [Enter]; then continue the Line to the Intersection of the entrance steps and the ramp wall. Erase the horizontal, vertical, and 3-foot line.

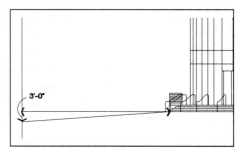

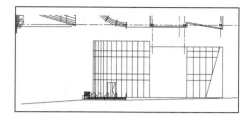

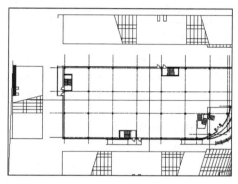

[image] Repeat the process on the south side of the elevation, as shown in the previous illustration, and EXTEND the south wall to the ground line. This is how the finished elevation should look. We now need to make a BLOCK of this drawing and write it out to a file named West_elev, just as we did for the North_elev.dwg.

Continue to ROTATE the plan and to draw elevations. As each elevation is created, make it into a Block using the [image] Make A Block tool, and write it to a file on disk using the WBLOCK command. When you are finished, the drawing file will look something like our example shown here.

Managing and Assembling Elevation Drawings

The concept of Production File Management promoted by this book is based on the idea that the maximum number of files (drawings) should be available to the maximum number of people working on the project at any time in the process of getting the building drawn. Creating files for each elevation, complete with their own annotations and notes, allows more people to have access to drawings at a critical time in the production process.

This is where the Paper space model begins to fall apart. If all the elevations are in a single file (which is necessary for Pspace view creation), only one person can access that file at a time. If changes are required to two or more elevations and the plans simultaneously, only one person is allowed access to drawings that could be worked on by several people in a manual drafting process, or our model. When the CAD system impedes rather than accelerates the production of drawings, it is difficult to justify using it.

Our model uses Reference Drawings (xrefs in ACADspeak), rather than Paper space, to assemble elevation sheets. This allows for maximum flexibility for production scheduling of changes and last-minute error corrections. It's not perfect, because drafting in AutoCAD or any CAD system is not a perfect process, but it gives you the best chance of bringing cyberdigits to paper on schedule and under budget.

Setting up the Elevation Drawing Sheets Files

Start AutoCAD. Open the A-1.dwt file by selecting the "Use a Template" button on the Start Up dialog, or select the same option on the File menu. Use the File menu to select Save As, and save the drawing with your Elevation sheet designation, for example, "A-101-1_Elevations.dwg".

Now select the XREF tool (XREF) on the XREF tool box (or on the Main tool bar if you followed our tool bar customization suggestions). The XREF dialog will appear. Pick the Attach button. Pick the Browse button and use the Explorer style directory and file display to find the North_Elev.dwg file.

Move the drawing into position on the screen as shown in the illustration. Now use the xref tool again to attach the Title_Block.dwg file we created in Chapter 4, using an insert point of 0,0. This will help us position the other elevation drawings on the screen relative to the positions they will be plotted in.

Continue to insert the other three elevation drawings as xrefs. You can now add titles to the elevations and the sheet number to the Title Block, and any other annotation or notes that you want.

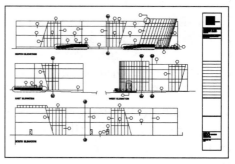

There are two schools of thought on where to put finish notes, detail and section cut keys, dimensions, and other annotation. One school puts all this information on the individual elevation drawing files; the other school puts it on the Elevation Sheet file, on top of the referenced drawings. Both have their advantages and disadvantages.

If you are drawing a large number of elevations (say, for a complex of buildings, such as a school or industrial park), you will perhaps need to have several people drawing and annotating elevations at the same time.

Putting annotation on each file will speed up this process, since more than one person can work on elevations that are referenced by the same sheet file. Under severe time pressure, this method may be the only way to meet the production schedule.

On the other hand, putting all the annotation on the sheet file makes formatting the final drawing much easier, and you avoid the possibility of information duplication. There is always the danger that a critical detail will get added to the sheet file at the last minute, and then several days later, added to the elevation file as well, when someone discovers that it is apparently "missing" from the elevation drawing. Having the annotation in one file also makes it easier to change things, because you don't have to open all the individual drawing files to add a detail key or note.

With all the keys and notes on the sheet file, the possibility of one of the referenced drawings growing too large to fit on the sheet (as notes and keys are added) is diminished, or at least more easily managed. Choosing which method to use is not a religious issue. What will work for one project may not work for another. Personalities may also dictate methods. Getting several architects to work collaboratively on drawings can be like herding cats, so the drawing process and structure may need to be designed to impose discipline. You will have to draw your own conclusion on this issue (pun unfortunately intended).

CHAPTER FOURTEEN

FINAL DESIGN DEVELOPMENT FOR PLAN DRAWINGS

Contents

In this chapter we will make final planning decisions regarding the building and add the final pieces of the puzzle to the first and second floors. Since our example building is intended to be a speculative office building, we wanted to use a single central interior corridor for exiting and access to the rentable spaces. This meant that all the spaces created by corridors leading to exits had to be truly rentable; that is, they had to be divisible by one or more minimum suite sizes, which in this case meant +/- 600 square feet. Once a test corridor layout was done, it became very clear that on the second floor, the distance from the southwest corner of the building to the north exit stair was over 150 feet. We would need to add sprinklers to the building, restrict all suites in the southwest bays to 300 square feet, or add an exit stair. In light of our goal of making the maximum amount of space easily leasable, the addition of an exit stair didn't seem like such a big deal. In a real project, this problem would have been solved in design development. We use it here as an example of how to use CAD to cope with last minute changes affecting more than one floor.

CAD Makes Major Changes Easy

Adding an exit stair or other major building element is relatively simple on a CAD system *if* it can be done prior to the creation of all the dependent plans needed for construction drawings. *An analysis of the design's functionality should always be made before beginning construction drawings.* I will always remember how the firm I once worked for proved to our client (the building's developer) that its landmark office tower, designed by a Big Name Architect, was not functional. The design was well into the construction-drawing phase when we demonstrated that the building would contain large areas that would be difficult or impossible to lease.

We proved this by drawing the plans on AutoCAD 11 and modeling the spaces created by the required exit corridors in ACAD 11's rudimentary 3D. It was easy to see how the maze of corridors required for exiting would eat up a significant portion of the rentable area on each floor. Not only that, but the corridors and exit locations would prevent a half or whole floor's being leased to a single large tenant. Since this building was to have a number of floors dedicated as a headquarters for a major financial institution, this was a serious flaw. The developer forced the Big Name Architect to go back and redesign the building, and that meant redrawing most of the construction drawings.

Adding the Exit Stair

First we need to create a COPY of the existing stair and move it to the location of the new one. We could insert our STAIR_PLAN1 block that we created in Chapter 12, but its insert point is at the door opening, which we would need to draw first, so in this case, use of the COPY tool is faster and easier. As the illustration shows, we use the Intersections of the column grid and the outside wall line as our COPY from and COPY to points.

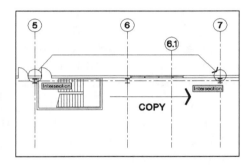

The next operation is to MIRROR the copy of the stair and then cut the door opening using the TRIM tool. The stair must be MIRRORed because we need to get the exit door on the second floor as far away from the west end of the building as possible. We use the **Midpoint** of the stairwell wall as our starting MIRROR line point and, of course with Ortho Mode on, create a vertical axis for the operation.

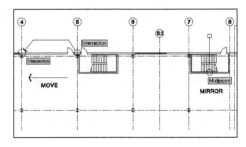

Now all we have to do is MOVE the existing stair to its new location. Again we use the **Intersections** of the outside building wall line and the column grid as our From and To points. All that remains is using the TRIM and FILLET tools to create a new door opening and to close the old one.

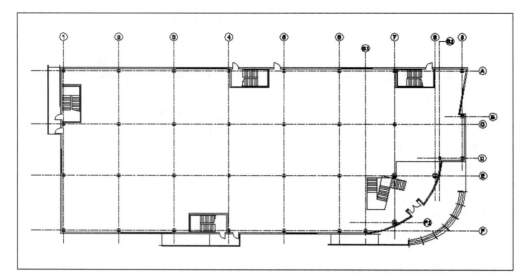

So here's our final first-floor shell plan. You have probably noticed that we have added letter and number identifiers to the column grid. These actually can and should be added earlier in the process. They definitely MUST be added before other plans are created from this base drawing.

We need to do one last thing, and that is to update the exit doors on the south elevation. Turn on the south elevation's layers (if you had previously turned them off) and draw 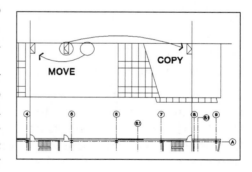 Construction Lines from the jambs of the new door locations on the south wall of the plan to the Elevation. Again, we COPY the old door to its new location, then MOVE the original door to the east, to its new location. The last step is to draw the hinge side indicator lines on the doors.

 To preserve the changes we have made to the Elevation, we will need to redefine the South_Elevation Block and write it out to the *same* South_Elevation file we just inserted. The first step is to ROTATE the plan 180 degrees so the elevation is right side up. Use the Block tool (BMAKE/BLOCK) to redefine (recreate) the block, and make sure to pick the SAME INSERTION POINT as before. Once the Elevation Block is stored in its South_Elevation file again, you can ERASE it. SAVE the Drawing! Use the file name A-1_base.

NOTE: If you forget an insertion point, just INSERT another copy of the old elevation block and note the insert point, then erase the block, or [U]ndo the insertion.

It is time to SAVE the first-floor plan and then use it to build the second-floor and roof plans. To start, save it again using the File Menu's Save As option, using a different file name, such as A-2.dwg, or whatever strikes your fancy or follows your firm's practice standards.

Updating the Second-Floor Windows

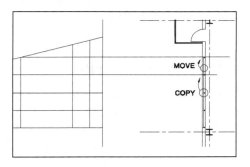

The windows are the next things to get our attention. We turned on the elevation's layers and projected the necessary lines from the second-floor level to the plan, and used the ⌐⁄ EXTEND, ✛ MOVE, and ▧ COPY tools to make the plan match the elevation's glazing width. This illustration shows the process with the east elevation and east window.

First, guide lines were drawn from the second-floor Elevation's key points and between the Midpoints of jambs on the plan. The end mullion and jamb were ✛ MOVEd to their new location, then a mullion was ▧ COPIed to the intersection of the window center guide line. The last operation required is to ▱ ERASE the guide lines. This process was repeated for each window.

Erasing All First-Floor Objects and Elevations

▱ The next step is to ERASE the first-floor exit doors (not the stair doors), close their openings, and erase the exit stairs themselves. It is also time to ERASE exterior ramps and stairs, property lines, etc.

Inserting Second-Floor Stairs

▨ The STAIR_2 block, which we created in Chapter 12, is inserted using the Insert Block (INSERT) tool that you used in earlier chapters.

All that is necessary is to drag the block to the Intersection of the interior building wall line and the exit door jamb. Remember that the southwest stair must be mirrored. In the illustration, the first-floor stair is shown as a dashed line reference, though it has been erased.

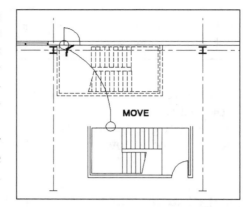

With the stairs INSERTED, we can now ERASE all the exit doors and jambs in the building wall, except for the lobby doors (we'll deal with them in a moment).

Drawing the Corridor

The main corridor of the floor is placed in the center of the floor's long axis to provide reasonably deep spaces on each side. Its placement leaves us with space that is easy to divide into 600- to 1200-square-foot suites. The southwest corner area, at +1300 square feet, can become two suites or a single suite. If a single tenant takes the entire corner plus additional space, then the exit corridor to the stair would be entirely contained within the tenant's area since its purpose is to provide a second exit for the occupants in this part of the building.

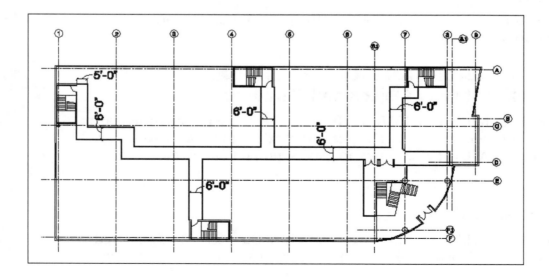

The central corridor is drawn on the Bldgwall layer because it is a permanent part of the permitted structure and must have a one-hour rating. A pair of double doors on magnetic hold-opens makes the lobby space at the stair landing feel open. We will add fire-rated glazing on the south and east sides when we complete the atrium. Almost all the corridors leading to exit stairs are the same 6'-0" width as the main corridor because they are potential access ways to suite entrances. With the corridor completed, we can now do the last part of the design work, which is locating the rest of the building core elements: the rest rooms and duct shafts and the elevator, equipment, and elevator machine rooms.

Completing the Building Core

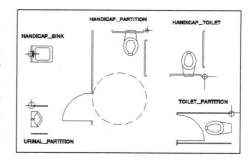

Adding the rest rooms is the next chore. There's no way to do this quickly, except to sketch out possible locations on paper first, and once the location and approximate size are decided, do test layouts on the CAD system. The same goes for the elevator and elevator equipment room, janitor's room, etc.

Planning and Drawing Rest Rooms

I have supplied you with a series of blocks (on the companion disk) to use in constructing rest room plans. This kit of parts will allow you to do test layouts of different toilet stall configurations, sink locations, etc. Here's an illustration of the blocks included on the disk: The insertion points of each block are indicated with a circle and right-angle lines drawn from the center of the circle.

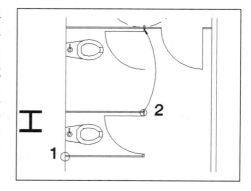

The TOILET_PARTITION blocks are designed to be inserted at the adjacent partition, or wall corner. Once you have one partition inserted, EXPLODE the block using the EXPLODE tool on the Modify tool box, then use the COPY Objects tool to duplicate the parts you need for subsequent stalls. In the example illustrated here, I first Inserted the disabled access stall, then developed the other stall in the following steps:

1. I used the Insert Block tool on the Draw tool box to insert the TOILET_PARTITION block using the **Nearest** Osnap.

2. I then picked the COPY tool on the Modify tool box and COPIed the toilet, door, and partition end piece, starting with the **Intersection** of the partition and end piece and dragging it **Perpendicular** to the disabled access stall. Once it was in position, I just use the TRIM tool to cut the end piece off at the disabled access stall's outside line.

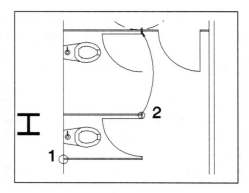

3. The last step is to MOVE the original partition and toilet into position by picking the outside **Intersection** of the partition end piece and dragging it to the Endpoint of the door arc above.

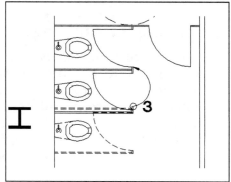

Now we can MIRROR the toilet stalls to the other side of the plumbing chase, thus easily starting the men's rest room layout. Sinks were inserted using Tracking: When you are prompted for an insert point, pick the Tracking button, pick the intersection of the rest room and baffle wall, drag the sink up (90 degrees), type 15, and press the Enter button on the mouse. Press the Enter button again to accept the default scale, and rotate the sink to the desired angle.

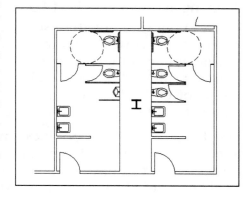

Completing the Lobby Atrium

The last bit of design development work in the atrium involves the lean-to geometry of the recessed entry in the curved curtain wall. There are a number of ways we could treat the entry, but we've chosen to go with the slant to the right (as seen from outside the building) and carry it through into the vestibule. This means our original idea of vestibule walls at 90 degrees to the floor plane that we illustrated in the lobby section drawing will have to change.

Finishing the Plan of the Entry Vestibule

1. The first step is to determine where in the plan the top of the vestibule will intersect the curtain wall. We do this by drawing a simple diagram as illustrated here—a triangle started with LINEs snapped to the **Endpoints** of the curtain wall mullions.

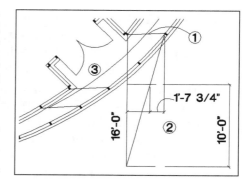

2. The 16-foot side is the floor to floor height. The long angled side is a section view of the back edge of the mullion. A LINE drawn 10 feet above the schematic floor line gives us the offset between the second-floor level and the top of the vestibule: 1'-7 3/4".

3. The inside curtain wall line is now OFFSET 1'-7 3/4" from its second-floor position. We draw two other mullion lines on the left side of the doors and EXTEND the arc of the vestibule roof to the leftmost mullion center.

Now we just need to draw four LINES to create the vestibule roof, and then we can erase the first-floor curtain wall and doors. Since the vestibule roof is below the floor plane, we draw it as a Hidden line type and do the same for the connecting mullions.

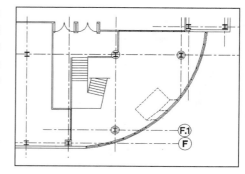

The final touch needed in the atrium space is to draw a cut line in the lobby stair and ERASE the lower portions. This completes our second-floor plan development exercise.

On a real project, we would make a copy (a block written to disk) of the building core and insert it on the first-floor plan, and then of course we would need to draw the interior corridors and other fixed elements on that floor as well.

Setting Up the Roof Plan Drawing

1. Turn off Bldgroof, Bldgcolumn, and Bldgrid layers. Erase all windows, the atrium curtain wall, and all interior construction (except the stairs and janitor room).

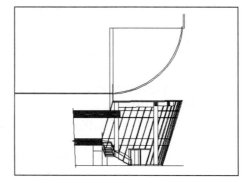

2. Turn on the Bldgroof layer and make it the current layer. Insert the building section drawing and align it with the end of the curve of the curtain wall. Project the slope of the clerestory windows and draw a 3'-0" x 4'-0" roof access hatch in the corner of the janitor room next to the leftmost south side stair.

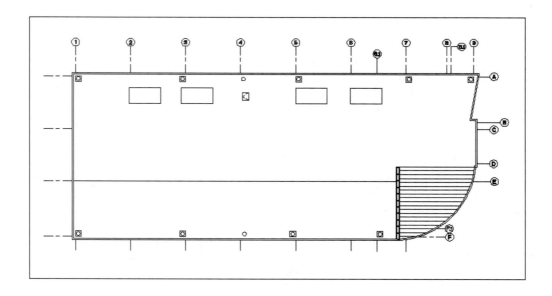

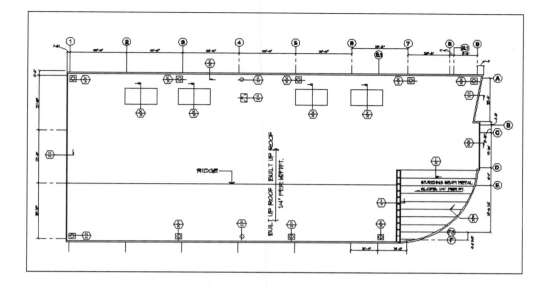

3. TRIM the column grid lines out of the roof, and draw the ridge
   line on column line E. Erase the stairs and the janitor room. Now
   we can locate the roof drains, sump pans, and overflow drains.
   Finally we add the level pads for the air-conditioning units and
   the standing seam pattern for the metal roof over the atrium. The
   standing seam pattern was created by first BREAKing the upper
   clerestory line where the east window slope begins, then using the
   Rectangular ARRAY tool on the Modify tool box to create the 19
   rows at -2'-0" apart.

All that's required to turn this plan into a final construction drawing is
to add dimensions and slope annotation, roof type indications, and
detail keys once the engineers have verified the roof drain quantity
and location and have located the HVAC shaft penetrations. Later we
will overlay this plan on our second- and first-floor plans to finalize
the roof drain penetrations through those floors. In the meantime, we
can still add preliminary detail keys. The next chapter explores how to
do this in greater detail. Once the first-floor plan is dimensioned, we
can use an xref overlay to bring in the building perimeter dimensions,
as illustrated here.

However, we are getting ahead of ourselves: The next chapter will deal with setting up dimension styles, dimensioning plan drawings, and the religious and technical issues involving detail keys and other annotation.

CONSTRUCTION DRAWINGS

PART THREE OVERVIEW

CONSTRUCTION DRAWINGS

Contents

With design development complete, we are now going to relatively quickly draw and assemble the construction drawings. I am not trivializing the process or the amount of work involved, but a large part of the work is already complete because the design development drawings contain the skeleton of our final C.D. set.

Setting Up the Base Building Drawing Files

We created a first-floor base building plan the same way we made the second-floor plan in the previous chapter, and we saved it under the file name **A-1_base**. Next we opened up a blank drawing based on our blank A-1.dwg file that we first created for this project. We will use the XREF tool to Attach the first-floor A-1_base.dwg file plan to this drawing file.

Using Reference Drawings to Create Construction Drawing Files

 1. To create the Floor Plan, we will make use of our first reference drawing attachment. Pick the XREF tool to open up the External Reference Manager dialog box. Pick the **Attach** button: The File To Attach dialog box will appear.

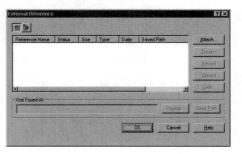

2. Pick the folder navigation buttons until you are in your project folder, and double click the A-1_base.dwg plan file we created earlier (alternatively, you can pick the file name and pick the **Open** button).

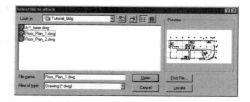

3. If the Base Plan's name is too long to attach to the current file, AutoCAD will politely suggest that we call it "Xref1" as an alias. This doesn't change the name of the file, just the Block Reference name inside Auto-CAD's drawing database. You can accept the suggested name or type in your own short name, such as "Base1".

4. Now the Attach Xref dialog will appear with some more choices for you to consider. The most important one is the Include Path check box. Leave it unchecked if the reference drawing resides in the same folder as the drawing you are working on. *Always* leave it unchecked when you are sharing files containing xrefs with consultants. If you have the referenced file in a folder other than the one you are drawing in, make

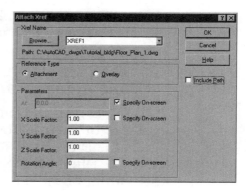

sure the box is checked so that you will not have to manually reattach the xref every time you open the drawing.

The Overlay option is a nifty way to get around AutoCAD's refusal to allow circular references. Circular references occur when two drawings in a reference group try to call the same reference drawing. Recently I needed to open a Furniture/Equipment Plan that called the Base Building Plan as an xref. But I also needed to xref in the Floor Covering Plan which called the same Base Building file. The solution was to open the Furniture/Equipment Plan and **Overlay** the Floor Covering Plan on top of it.

The Parameters can be largely ignored unless you get into sophisticated xref scaling schemes. We don't even play with these in setting up detail sheets, if at all possible. The most important button to pick is the OK button to get on with completing your reference drawing insertion.

Note: When preparing files to transfer to your consultants, if you have the path included in your xrefs, you will need to go back to each drawing and reattach the reference files with the Include Path box unchecked. The drawing file calling the xrefs must then be saved to preserve the new setting.

Pick the OK button and the External Reference Manager dialog box will disappear, and you will be prompted for an insertion point on the command line. Type <u>0,0</u> and press [Enter]. You should now have a drawing with the Base Building Plan inserted into it.

Using the **Save As** option on the File Menu, we save this hybrid drawing as the A-1_floor_plan.dwg file, and we're ready to start our first construction drawing: the final floor plan of the first floor.

Setting Up Dimensioning Parameters for Plan Drawings

 1. Pick the Dimension button on the tool bar, or use the Tools Menu to access the Dimension Tool Box, and pick the Dimension Style tool on the bottom of the box. The Dimension Styles dialog box will pop up on the screen as shown in the illustration.

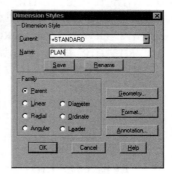

 2. Type <u>PLAN</u> in the Name window and pick the Save button under the window. This creates a new dimension style within the drawing file.

3. Pick the **Geometry** button (or press [G]). The dialog box for setting all the parameters of the dimension and extension lines will now be displayed. Set the parameters as shown in the illustration. We have selected Architectural Tick as the

arrowhead style because it reads better on plotted drawings, but beware that it is not supported by Release 12 or 13 and therefore may not translate properly. Type the values illustrated for the Dimension and Extension Lines. Pick the OK button when you have completed entering all the parameters.

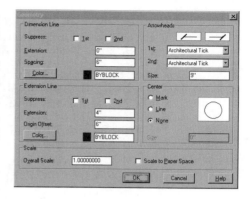

4. Now pick the Format button (or press [F]) on the Dimension Style dialog box. Use the check boxes and drop-down lists in the Format dialog to set up parameters for the relationship between the dimension text and dimension/extension lines. Selecting the No Leader option will enable you to move dimension text that is too big to fit between extension lines so that you can add a leader from the dimension line to the text. I personally find this is much faster than using AutoCAD's leader routine, and I can exercise better control over the appearance of the drawing. Select the options for justification and text alignment, as illustrated. Pick the OK button to return to the Style dialog.

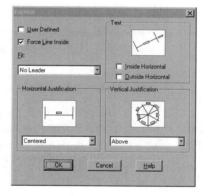

5. Now pick the Annotation button (or press [A]). Press the Units button or the [U] key. The Units dialog will appear over the Annotation dialog box. You can ignore the Tolerance settings (some architects are famous for a lack of tolerance anyway). From the Units drop-down list, pick Architectural, not Architectural (stacked). This latter style is hard to

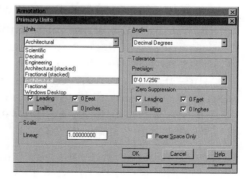

read in plots under ½" = 1'-0" scale, and it takes up more vertical space without returning any real horizontal space benefits. Set the rest of the parameters as shown, and pick OK to return to the Annotation dialog. Note that we set Round Off value to ½". This is more fine than can actually be built. It should be set to ⅛" or ¹⁄₁₆" for detail drawings.

6. Finish the annotation settings as shown in the illustration. The text Gap is the distance above the dimension line that the text will be located. Make sure to set the Round Off value to the same ½" as the Precision setting. Pick your favorite text style, type in the height of <u>10"</u> for the text, pick the color (if you want it different from the dimension and extension lines), and pick the OK button.

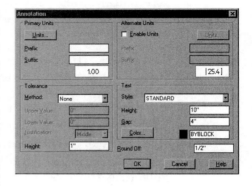

As a general rule, I pick a color for the dimension text that is mapped to a slightly heavier pen weight and use a light line for the dimension line and extension lines. Now we are back at the Dimension Styles dialog box. Pick the Save button one more time, then pick OK. The settings you just entered will produce dimensions similar to the ones illustrated here, depending on which Text Style you use.

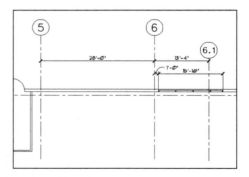

Setting Up Standard Dimension Styles for All Types of Drawings

AutoCAD does not store dimension styles in separate files that a variety of drawings can access (another feature); rather, it stores dimension styles as part of the drawing file. This forces us to create template drawing files incorporating dimension styles for plans, details, large-scale plans, etc. Happily, Release 14 makes dimension style setup much easier and more flexible than earlier versions, and you only need to do the setup once, then save the file(s).

Dimension styles need to be created where you will be plotting Model Space drawings at defined scales. For instance, in a ¼" = 1'-0" scale plot, you will want the text height set to 5" instead of 10", and the tick marks should be half the size of the ⅛" to 1'-0" scale style.

Dimensioning the Floor Plan

Now we are going to proceed to dimension the rest of the first-floor Floor Plan and add building and detail section keys to it. The first step is to create two dimension layers: one for the structural grid and one for other building dimensions. The structural grid layer might be named "Grid_dimension" and the building layer "Bldgdimension". We will start by making Grid_dimension the current layer.

Start by setting your running Osnap to **Intersection**, **Perpendicular**, and **Nearest**.

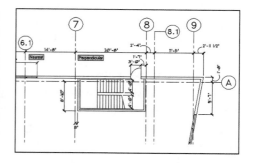

Dimension the building grid by picking the Linear Dimension (DIMLINEAR) tool on the Dimension tool box, then picking a grid line and picking the next grid line. Autosnap will automatically "know" that you are selecting Nearest to Perpendicular. Work fairly well zoomed in, as illustrated, pressing the mouse's Enter button to recall the DIMLINEAR command.

Release 12 users will find this very fast and will be relieved that they no longer have to type in "HOR" (horizontal) or "VER" (vertical) to set the dimension mode. As you work around the building parameter, dimension key wall and stair offsets and other critical items. Don't forget to change layers as required, or you will pay for it later (as you will see when we draw and dimension the Foundation Plan).

This illustration shows the entrance lobby area completely dimensioned. You will note that the stair has no dimensions: It will be dimensioned and detailed on a separate drawing, along with the fire exit stairs. Even though we have some dimensions yet to add to the north side of the building and the interior, we are going to pause here and start adding detail keys, room numbers, and door numbers.

Detail Keys, and Room and Door Numbers

The symbols used for drawing annotation are largely a personal and/or religious issue. As far as CAD drafting is concerned though, there are some shapes that are more difficult to draw and some key designs that are more consuming of system resources than others. The following illustration shows some of the types of annotation symbols used in our office's drawings.

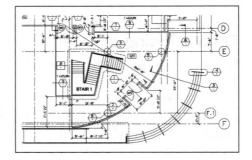

All text is inserted into symbols using the ▣ DTEXT tool with justification set to Middle. Middle justification is set by pressing the [M] key before you pick a start point for the text. When you pick the start point of the text, pick the visual center of symbol or space within the symbol that the text will occupy.

Drawing Detail Keys

Specific details are keyed with a hexagon that is divided in half with a horizontal line. The detail number is above the line, with the sheet number below. At this stage, sheet numbers are not assigned, so we have just put a hyphen character here. As we assign sheet numbers, we will go back and use the ▣ Edit Text tool to change the hyphen.

Details of parts of the building such as stairs and rest rooms that are going to be dimensioned on larger scale plans, elevations, and sections are indicated by drawing a dashed line (Hidden line type) around the area and keying it with a square symbol. Using a different symbol is just a personal preference for me. I feel it makes finding the detail key easier amid the visual clutter of dimension, notes, and hex key symbols.

Both the square and hex symbols are created using the ▣ Polygon (POL) tool. In this drawing, they are 3'-0" x 3'-0", circumscribed around a 3'-0" diameter circle. Of course, you are free to use whatever symbol shapes you are comfortable with, assuming you have a choice.

Direction Indicators

There are many ways to indicate the direction a detail section or building section is cut. Here is the symbol (on the left in the illustration) commonly used in manual drafting, compared with the one more suited to CAD drafting.

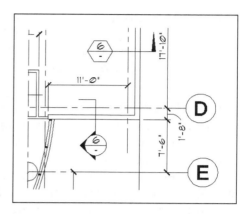

The lower symbol was created by drawing a square POLYGON around a CIRCLE and then rotating it 45 degrees, TRIMming it to a vertical line, and then HATCHing the dark areas with Release 14's new Solid hatch type. Release 14 makes drawing this type of graphic fairly easy. The other symbol is a hexagon created with the POLYGON tool. It uses a Polyline (PLINE) with a zero width at its tip and a 6-inch width at its base as an arrow to show direction of view. This symbol even in Release 14 has much less overhead (requires fewer system resources) than the more traditional one. There are additional overhead problems with the traditional graphic related to file size and both the Release 12 and 14 file formats.

When the drawing is SAVEd to Release 12 format, the solid **Hatch** is converted to a **Solid**, and the drawing file size increases dramatically. For example, the same drawing shown in the illustration saved to Release 12 format with only one of these symbols is 262.1 kilobytes in file size. When the symbol count is increased to 25, the file size balloons to 290.8 kilobytes, a 28.7-kilobyte jump.

Naturally, there's always some increase in file size as objects are added to the drawing, and adding 25 hex symbols in Release 12 adds 10.7 K to the file size, 17.9 K less than the manual drafting style circle and arrow. In Release 14 format there is a smaller 12-kilobyte penalty for using the circle and arrow, due to R 14's more efficient handling of hatches.

I bring up this subject to illustrate how your choice of graphics can penalize your CAD system. If you have large numbers of complex objects and lots of hatches, you will get drawings that are huge in terms of file size and that take forever to load, save, and regenerate. **What you draw** can affect the performance of the system, and therefore its overall productivity. I try to use symbols that communicate well but that are minimally stressful for the CAD system.

The next illustration shows a way to indicate a specific typical detail, or a small area section where the direction of view is not important (because the object being sectioned is symmetrical). I simply draw a 3-

inch-wide Polyline over the area being detailed, to emphasize the restricted nature of the objects involved. In this case, the detail is for magnetic door hold-opens.

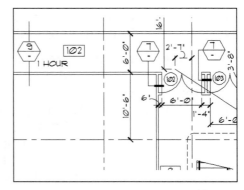

Wall Type Indication

This same illustration shows a simple way of indicating wall types. Many people use hatches to indicate wall types. I have always considered doing so wasteful of time and energy in manual drafting, and I consider it wasteful of system resources in CAD drafting, even with Release 14's improvements in handling hatches. Here is my alternative to hatching or other types of embedded graphic indication of wall type: a detail key with the wall type rating (if it has one) placed next to it.

Our experience with most building department plan checkers is that they like this method and it saves having to put a wall type legend on every floor plan. Legends are another error-prone area of drafting, similar to schedules, and make on-screen checking difficult, since the information they represent may not be visible when you are zoomed in enough to see what type of hatch pattern you are trying to verify. You can use viewports to overcome some of this difficulty, but you must work in a smaller area of the plan within the restricted screen window of the viewport.

Room Numbers

Room numbers are quite simple: We use a rectangle 3'-0" x 2'-0" with 1'-0" text inside it. If your firm has a standard, use it. We just try to make this an easy-to-identify shape, different from door numbers and detail keys.

Door Numbers

In this illustration, I have used the CSI standard of sequential and individual numbering, which I am not a great believer in, especially in this era of CAD document exchange with general contractors and architects/designers. **Whenever possible, we try to number doors by type instead**. For example, in our practice, all doors numbered 102 would be 20-minute rated, in hollow metal 20-minute-rated frames, stain-grade maple veneer, with a particular latch and lock set. There may be 105 or

107 of these doors on the drawings, but we do not provide a quantity total to the general contractor (we do count doors for budgeting and cost analysis, but those numbers are never on the drawings). My main concern is to put the correct door/finish/hardware combination in the correct location.

Contract Issues Regarding Drawing Schedules

We feel it is the responsibility of the general contractor and the subcontractors to do an accurate count of doors, just as with every other building component on the project. We do not want to put our firm in the position of being liable for counting or cataloging building components, because in today's world, we can't get paid for it. Think about it: You don't individually number every foundation footing for every column, do you? Why follow a different procedure for doors or other building components? I can hear the immediate chorus of voices saying "But how can we check the contractor's takeoff?" The answer is: **You don't**. Do you have an on-site project architect checking the number of wall studs? If the answer is no, then don't spend fee money checking something for which someone else should be liable and should be getting paid to control.

Polemics aside, design your door schedule symbol so that it cannot be visually confused with other shapes on the same drawing. This is especially important because doors tend to be the focus of other detail indications, both in plans and in elevations.

Building Section Indicators

As you can see in the illustration, we recommend a simple square symbol with a filled triangle on the view direction side. The square shape provides the most room for text, and that's the reason we use it. The filled triangle is just a Polyline with a starting width the width of the square (4'-0" in this case) and an ending width of zero. Polylines are used to create the view direction arrows at the end of the section lines, just as in the other detail symbols.

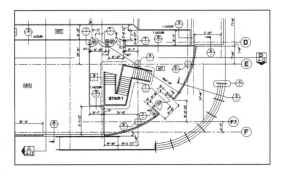

If you prefer to use the more traditional circle and arrow, go ahead. The overhead cost to the drawing will be minimal because few building section indicators are needed on either elevation or plan drawings.

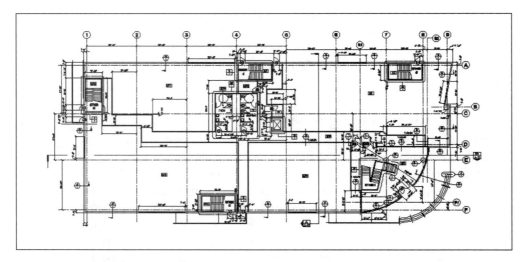

This last illustration shows the complete floor plan with its dimensions and section lines, details, and other annotation.

Note that we have gone back to a door-numbering system different from the CSI standard. In this system all first-floor doors get a type indicator letter (A, B, C, etc.) followed by the number 1. In the door schedule, doors are listed by floor, in alphabetical order.

Before we start the Foundation Plan, we need to make a Block of our standard detail key symbol and write it out to the block library on disk, using the WBLOCK command. Give the block a name that indicates its intended use, such as PLAN_DETAIL_KEY. I like to use all uppercase letters for drawing components to distinguish them from other types of drawings. That way you can easily distinguish blocks from other drawing files when looking at the directory listing in AutoCAD's explorer style file window.

Drawing the Foundation Plan

Once you have saved the Floor Plan, Open the A-1_base file and use Save As to save it as the Foundation Plan, using your file name standards. The current version of the Base Plan becomes the geometric basis for the Foundation Plan, and then the actual Floor Plan file is attached to the Foundation Plan as a reference drawing once the Foundation Plan is completed.

Initial Drawing Setup

Turn off all layers except the Bldgwall layer, and erase all the interior partitions except the elevator, rest rooms, and stairs/stair shafts. Erase all the interior corridor walls and the rest room walls except where the plumbing chase is located. Create a new layer for foundations (we named ours Bldg_foundation), and give it the Hidden line type. Turn on the Bldgrid layer, but not the Bldgcolumn layer. Make the foundation layer the Current one. Now we are ready to draw some footings.

Column Pad Footings

On this building we are required to use 5'-0" x 5'-0" column pads under all the main structural columns. Use the Polygon tool to draw a 5-foot square centered on one of the grid line intersections. Use the COPY tool and its Multiple option to place the pads at all grid line intersections. Turn on the Bldgcolumn layer to make sure you have accounted for all the columns, and erase any pads accidentally placed on intersections without columns.

Drawing Wall Footings

Next use the Offset tool to create the footing for the plumbing chase support at grid line 4C, as shown in the illustration. Once you have the footing roughed out, use the Fillet tool to complete the corners and then use the Match Properties button on the tool bar to change its layer and line type. Erase the wall lines.

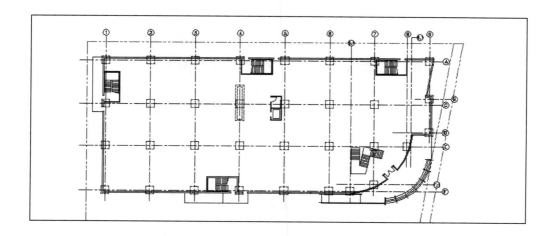

Repeat this procedure at all the exterior building walls, as well as the elevator shaft. Once you have a footing drawn, erase the wall lines when you are zoomed in enough to pick them easily. In this building, we OFFSET the footing lines 8 inches outside each wall line and then used the 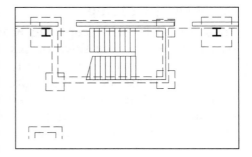 Match Properties tool to make them the correct layer and line type. The FILLET and TRIM tools were used to clean up intersections and corners.

Note that we could have used the LINE tool with Tracking to draw the footing lines, but in benchmarking that method, I discovered that there was a significant enough time savings in using OFFSET combined with the new Match Properties tool to make that the preferred procedure.

The next thing to do is to create and locate the 3'-0" x 3'-0" pads for the stair supports. These are again a 4-sided POLYGON whose center is located at the **Intersection** of the corner of the stairs' landings. We then use the TRIM tool to cut out the footing lines passing through the pads.

Proceed this way all the way around the plan. Note that you only need to OFFSET one section of the building perimeter wall on each major side. The FILLET tool will join everything nicely, and your last job will be to use the TRIM tool to cut all the footing lines inside the pads. Note that this task is much easier with the Bldgcolumn, Bldgwall, and Bldgrid layers turned off.

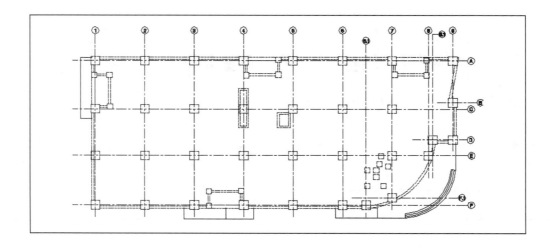

Dimensioning the Foundation Plan

Now we need to turn on the Bldgrid layer to begin our dimensioning exercise. The Dimension Style settings used for the Floor Plan are still current in this drawing, and we will use them as the default style for dimensioning the Foundation Plan. Now we can turn the Grid_dimension layer on, and we will start with the building grid already dimensioned. Widths of footings will be dimensioned on the detail sections. What we want to dimension here are the relationships of the footings to the structural grid.

Set the Running Osnap to **Midpoint** and **Intersection** and **Perpendicular** only. Start dimensioning from the Midpoints of the column pads Perpendicular to grid lines. Continue to dimension all the key relationships in an area at a comfortable zoom factor, then ZOOM Previous and move to another area.

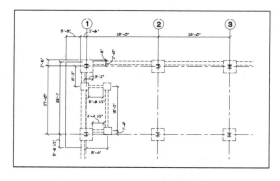

Areas that are too congested for annotation and dimensions to coexist will be duplicated in another file and annotated there. We have chosen this method for the entrance lobby area, just as we did on the Floor Plan.

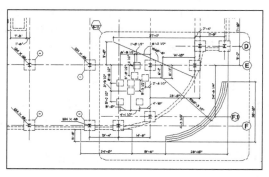

Annotating the Foundation Plan

Column pads are dimensioned in a schedule located at the lower side of the drawing. As the illustration shows, we have placed a circle with a line at a 45-degree angle to the column pad to create the schedule key symbol.

The symbol was easily created by first drawing a CIRCLE with a radius of 15", then drawing a short LINE from the circle's right **Quadrant** Osnap zero degrees to about 3' away from the circle. The circle and line were then ROTATEd 45 degrees and moved from the line's **Endpoint** to the column pad's **Intersection**.

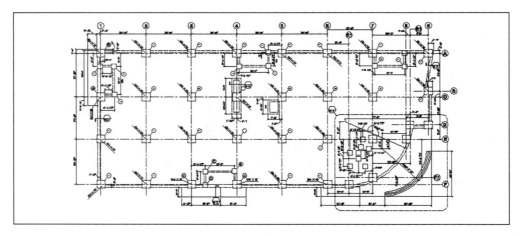

We need pad key symbols in other orientations too, so we just MIR-RORed the original in four different directions and copied them throughout the plan. You will note that the key symbol has only a dash in its center instead of a letter. This allows us to stay fairly zoomed out and still use the [A'] Edit Text (DDEDIT) tool to change the dash to the proper key reference after we have set up the pad schedule. The other bit of annotation placed on the column pad is the steel section note, which was created much the same way as the pad key, by drawing a line under the text, then rotating both 45 degrees and snapping the line's **End-point** to the Intersection of the grid lines. We selected the most common size column for the text; that way we can get away with less editing.

This illustration shows the pad keys in place and all the column annotation complete. We're now ready to make the column schedule. Then we will apply the letters, steel section text, and detail keys.

Making the Column Pad Schedule

[A] The pad schedule was typed using Auto-CAD's Multiline Text Editor tool (MTEXT) to start with, making sure to create a big enough box for everything to fit in. I set the text height to 1'-2" and used the space bar to separate the columns, typing only the body of the schedule. I left a blank line between rows of text.

COLUMN PAD SCHEDULE			
SYM.	SIZE	DEPTH	REINFORCEMENT
A	5'-Ø' SQ.	1'-4'	6 - #5 EW
B	5'-Ø" SQ.	1'-2'	4 - #5 EW
C	5'-Ø" SQ.	1'-Ø'	4 - #5 EW
D	3'-Ø' SQ.	1'-Ø"	4 - #5 EW

Next, the column headings were typed using the DTEXT tool. I increased the text height to 18 inches and typed the table title. The next step was to draw a ▱ RECTANGLE around the whole mess and draw the first three top horizontal lines using **Near** and **Perpendicular** Osnaps. Finally I used the ▨ COPY tool with the Multiple option and Ortho Mode on to place the last three lines. I snapped the first point of displacement as the **Intersection** of the second line from the top of the box and copied the third line down by successively snapping to the Intersection of each previously placed line.

Getting Rid of Nonfoundation Objects

It is time to completely convert the drawing file from a child of the Floor Plan to its own identity: Turn off all Foundation Plan layers and the Bldgrid and Grid_dimension layers, and turn all other layers on. Erase everything that is visible. Yes, I know we are erasing the ramps, columns, and entrance stairs, but we'll get them back—honest.

Using Purge to Clean Up the Plan File

Turn all layers back on and ◉ ZOOM into a quadrant of the drawing that requires detail annotation; INSERT the PLAN_DETAIL_KEY block we created earlier. In the illustration, we have started at the southeast stair. Once the block is inserted, we can PURGE the drawing. Purging at this point will get rid of all unused layers and block definitions left over from the Floor Plan.

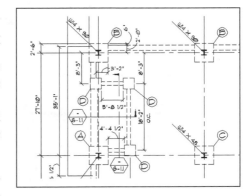

Purge tips: You can quickly get rid of unwanted blocks, line types, and layers by typing N when the Purge command prompts you to confirm all deletions (the default is to confirm that you want each item purged). Use this command with care. If you are not sure about what you are throwing out of the drawing, confirm the selection of each object. You should also run PURGE at least twice, because AutoCAD will throw out an unused block or text style but won't also dump the layers associated with the purged objects unless you run the command a second, or even third, time. It can be argued that this is a feature, not a bug.

Now we can complete the insertion of detail keys and use the Edit Text (DDEDIT) tool to modify the column and foundation pad annotations. Remember to insert the text in circular keys using the **Center** Osnap and set the DTEXT justification to Middle.

Working with Reference Drawings: Attaching the Floor Plan to the Foundation Drawing

To finish the foundation plan, we will make use of our next reference drawing attachment. Pick the XREF tool to open up the External Reference Manager dialog box. Pick the **Attach** button and use the Browse button just as we did at the start of the chapter to locate the A-1_base file.

When the External Reference Manager disappears, you will be prompted for an insertion point. Type 0,0 and press [Enter]. The Base Building Plan will now be inserted with perfect registration, "underlaid" beneath the foundation plan. Of course the result is a mess, as you can see in the illustration. We need to turn off all the Floor Plan's unwanted layers, but more than that, we want to make sure they are always automatically turned off every time we open the Foundation Plan drawing.

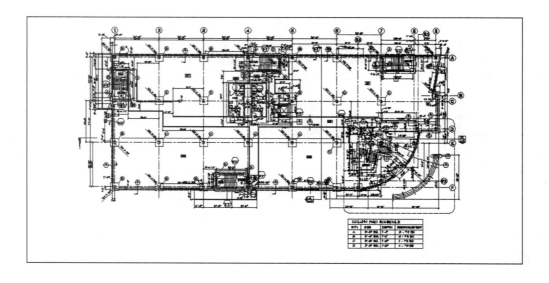

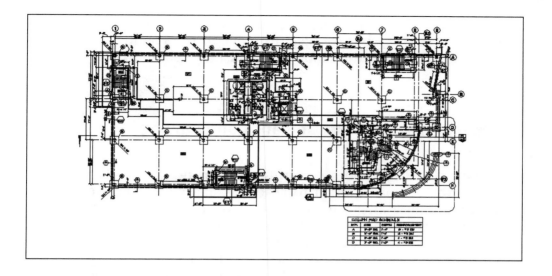

Using the VISRETAIN Variable Setting in Xref Drawings

On the command line, type <u>VISRETAIN</u> and press [Enter]. AutoCAD will parrot the Visretain variable name on the command line and prompt you for a new value. If the current value is zero, type <u>1</u> and press [Enter]. Setting Visretain to 1 will ensure that our layer settings will always remain the same once we SAVE the drawing. Now use the Layer dropdown list on the Object Properties tool bar to turn off the Floor Plan's dimension and annotation layers, stair layers, door layers, and all other layers duplicating objects on the Foundation Plan (now you see why we put the ramps and exterior stairs on a separate layer). The result should look like this illustration.

SAVE the Foundation Plan, and our work is complete. Yes, we could have left the building grid out of the foundation plan file, but if we work on the file without the Floor Plan Xref attached, it is useful to have the grid available. It doesn't make that much difference; it's really a drafter's choice. We duplicated the building grid dimensions because we needed to move some of them to make room for our foundation dimensions and annotation, especially on the east and west ends of the plan.

Since the grid and its dimensions are the least likely to change, we are not running much of a risk here. In the event that there are major changes to the building grid and its associated dimensions, we would simply erase them from our Foundation Plan drawing and turn on the layers in the Xref file, move what we need to move in the Foundation Plan, and at the worst, reopen the Floor Plan and move some dimensions there.

Reference Drawings Make New Demands on Drawing Managers

The use of reference drawings is both powerful and demanding. This is a drawing structure that requires much forethought at times when we normally are not thinking ahead to the next two or three drawings down the line. The biggest demand load is in setting up the layer separation of the base building drawings, since they underlay everything from structural drawings to reflected ceiling plans for interior design. The most important thing to remember is that a mistake in base building layer setup is never fatal. AutoCAD allows changes at any time. Just make sure that what you change on one floor is carried through to all the base building files on the project, as necessary.

For those of you working on 45-floor high-rise projects, I suggest making the building grid and dimensions a master file called by all floor plans as a reference drawing. Pick a "middle common denominator" part of the structure that has the most commonly shared objects with all other floors as the base reference file. The rest of the dimensions can be added in the local drawing files as required. This part of the drawing setup is best accomplished by working with the building design architect or by the design architect directly setting up the CAD file. It is not something you can shoot at in the dark or try to make "office standards" handle, because projects are just too different. A drawing structure that works for a high-rise will almost never work for a business campus of multiple low-rise mixed-use buildings or a shopping center.

CHAPTER SIXTEEN

BUILDING SECTIONS AND STRUCTURAL FRAMING DRAWINGS

Contents

281

These final drawings are built directly on the Design Development files we created earlier. The process from design to final drawing uses the precision of the CAD system to keep the focus on the end result, which hopefully can be supported by the structural engineer's calculations and the client's budget. Naturally, at this stage of the project, there will be accommodations to the demands of local code authorities as we go through preliminary plan check. Changes to the drawings completed to date are not disastrous, because we are just starting the final annotation and detailing process, where most of the time is invested. Of particular importance is that we have not even touched the most complex drawings of the interior of the building—electrical, reflected ceiling, FF&E, HVAC, and other detail and derivative drawings that depend on these files.

Starting Building Section Drawings from Elevations

In this example, we will take the East Elevation of the building and the Roof Plan to set the geometry for the building section. The principle works for all elevations/sections for any building but will need to be adjusted for the complexity of the design you are doing. We are going to start in reverse, using the East Elevation, then MIRRORing everything to match the direction of the section view.

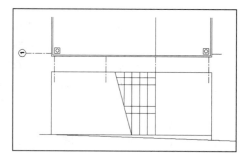

Open the East Elevation file, and Save As a new drawing file in the project directory. Attach the Roof Plan file as an xref and use the Rotate option to put it in relative orientation to the elevation. Use the MOVE tool to position it Perpendicular to the elevation's cardinal features. ERASE the elevation's exit doors and the guardrail.

Since the Roof Plan is an xref, we can't manip-ulate its objects (this is actually a good thing, as we will show), so we need to make a new layer for Buildgrid in this file and draw col-umn center lines from the Intersections of the building walls of the Roof Plan to the floor slab of the Elevation. If this is a long building section or an overly complex one, we would draw one line and use COPY with the <u>M</u>ultiple Option to make the lines.

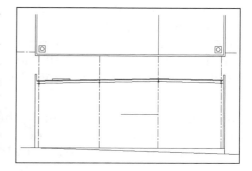

Setting the Roof Pitch in the Section

The design of the parapet at the deepest slope of the roof was originally set to 3'-0". Since our roof is asymmetrical, we will use this to set the slope for the longest run, letting the short side parapet be lower. This is a design, not a technical decision and it may be affected by local codes and all sorts of other considerations on a real building, but bear with me as we go through the process.

On the left side of the elevation, we draw a LINE from the upper left corner of the building 8" to the right, then down 3'-0" (270 degrees), using Direct Distance Entry. With the LINE tool still operative, we con-tinue 4'-0" to the right (0 degree), then up 1" (90 degrees), and finally back to Osnap to the **intersection** of the parapet wall and the 4'-0" line. Next we ERASE the 4'-0" horizontal and the 1" vertical line. Now we have a short line drawn from the base of the parapet at the correct slope of 1:48.

Using the EXTEND tool, we project the roof slope to the column center line that corresponds to the ridge line. We then MIRROR the roof line using the column line at the ridge as our Mirror line. Finally we use the FILLET tool to trim and join the parapet wall and roof lines. The final step is to project the air-conditioning equipment pads down to the elevation and draw a horizontal line using Tracking, 6" above the high slope side **Perpendicular** to the parapet side of the pad projection line. The TRIM tool takes care of creating the vertical sides of the pad.

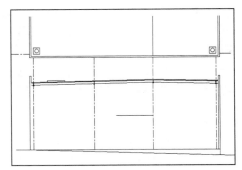

Now we use the ⬳ OFFSET tool to create the roof surface line and the frame members supporting it. We employ OFFSET again to create the building walls at the 8" design depth. TRIM and/or FILLET will be used later to clean up the intersections of wall and roof framing. We have also ⬰ ERASEd all the window wall lines except for the second-floor elevation line.

Note that we could have made the roof slope asymmetrical (a different slope on the short side of the ridge), keeping the parapet height the same on both sides of the building. Using this sort of design, however, means we will have a fractional value for the short side slope. This is really a designer's call. We will get a fractional value for the parapet height above the roof on the short side as we have now drawn it and in either case must use a +/– notation on these dimensions. Note that we are still making design decisions well into the construction drawing process.

Drawing the Second-Floor Section

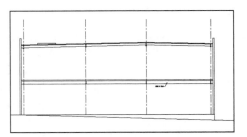

The next step is to use the PLINE tool to draw cross sections of the frame members that run east and west in the building. We EXTEND the floor elevation line and OFFSET the floor and its supporting beam line. We don't draw any detail of the metal floor deck or lightweight concrete on top of it, just as with the built-up roof system. That will be taken care of in detail drawings and annotation.

Drawing Smarter

This issue actually has nothing to do with CAD drafting except that implementing a CAD system gives everyone an opportunity to change their ways and become more efficient. We don't draw a tiny section of the floor deck, because it adds no meaningful information to the drawing. Instead, our note with its leader arrow pointing to the floor will convey precise information about the floor structure: the depth of the lightweight concrete on the particular type of metal deck.

For many years early in my career, I believed in the value of redundancy of visual and textual information: Draw it *and* note it, I believed, and the contractors will give you a better bid. Even hundreds of hours spent reading drawings to contractors over the phone did not dissuade me from that conviction. I did not change my mind until I learned about the disaster at the construction of a major corporate headquarters building in Colorado.

On this project, in spite of all the drawings, in spite of all the details and specifications, the contractor poured 80 percent of the floor slabs 1 inch thicker than the design called for. By the time the mistake was caught, only the lightly loaded executive floor was left unpoured. Because this building was to be occupied in the days of paper files (before personal computers), the effect of the mistake was even worse than it might be today, but no matter: The original general contractor went bankrupt (or was replaced—I'm not sure which), a new one was brought in, and new structural reinforcements had to be engineered and constructed at great cost and delay. I was told that even lighter systems furniture had to be found for the open plan areas.

Thousands of hours of unbillable (or uncollectable) time was spent in the process, and eventually several firms joint venturing on the massive project ceased to exist. The moral of this true story is that *drawings are not the product*. Yes, drawings need to be correct and complete, but there is a point at which there is no substitute for field observation. At that point, the drawings are just the *guide* for the architect to use to check for mistakes in the construction, because you can bet the contractor is not looking at them very closely.

Drawing Foundations and Floor Slab Sections

Again we use OFFSET to create the 6" floor slab, and we draw a schematic foundation section using the engineer's calculations and details. Note what we do not draw: We leave out rebar, reinforcing mesh, and doweling. We don't even hatch the slab and foundation with a concrete pattern, saving these for the details. At this stage we have also added reminder notes of the steel sections used in the roof framing, just so we don't lose track of this information.

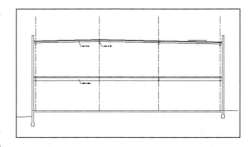

Column and Soil Indication

Using the 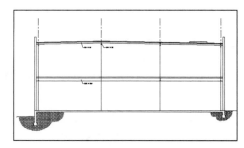 ARC tool with the 3-point setting, we draw a series of freehand curves as the
boundaries for the HATCH pattern for the
soil hatch. This is done for the benefit of permit application rather than construction, as
every contractor assumes that the building is
not supported by thin air. Most building
departments, however, want soil type information from the soils test keyed to something,
and this provides a graphic device for doing this if necessary. Once the
HATCH is applied, the boundary curves are erased, again just for graphic
style reasons. Your own taste rules here; just don't spend hours fiddling
with splines and other "artistic" shapes and HATCH scales. A HATCH
scale of +/– 3'-0" works pretty well. We have also drawn the column
frame lines as a Continuous line type from the underside **Intersection**
of the roof with the supporting wide flange beams with the Intersection
of the column center lines with the first-floor slab line.

Using the Floor Plan to Set Interior Walls in the Section

Here we insert the Floor Plan of the second or
first floor as an 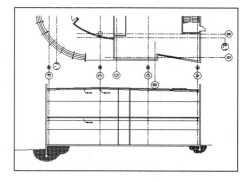 xref so we can project the
interior partitions cut by the section view. On
most projects, you will need to xref all floors,
one after the other, and project the plan objects
to the section view. The illustration shows how
this works with the First Floor Plan as we project
the central corridor into the section. Column
key bubbles are added as 2'-0" diameter circles
with 18"-high DETEXT letters middle-justi-
fied to the **Center** Osnap point of the circle.

Using the xref method in Release 14 is easy and quick because the dialog
box makes xref drawing management reasonably visual and gives the
user tight control over all the options. A bonus of using xrefs is that,
unlike inserting drawings as blocks, the layers of the xref are never
inserted into the current drawing file, along with text styles, dimension
styles, and all sorts of other detritus. Once the xref is Detached, all this
stuff vanishes. You can, of course, save the drawing with the xref
Attached and still not increase the file overhead.

Final Annotation of the Building Section

Once the corridor walls are drawn, we start adding notes about the corridor partition construction, ceiling type, and exterior wall construction and roof slope. Key building dimensions are introduced. Because we built this drawing from the previous building elevation, the dimension style was retained and easily reused. The notes can either be done in the [A] Multiline Text editor (MTEXT) or as [Al] Dynamic Text (DTEXT). I prefer DTEXT because I can see it on screen as it is typed, which gives me the opportunity to fit the note to the space available without a lot of fiddling with the note's format.

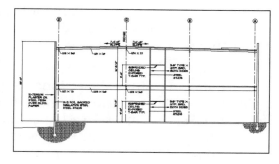

Final Drawing Output and Format

The DETEXT height was set at 9 inches to make it easily legible at a plotted scale of ⅛" = 1'-0", which will be the final output format for these building sections. As you can see, the drawing quickly fills up with notes. Wall support and framing details are now keyed in and will be located on the same sheets as the wall sections cut on the Elevations.

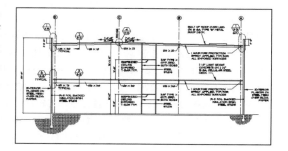

Building section drawings will be xrefed into a file that has just a title block (also xrefed) in it, with text added for drawing number, name, and dates/revisions, etc., just as we did for the building elevations. Section titles are also part of this file, not the individual drawing files being called as reference drawings. Why? Because as a general rule, if it becomes necessary to add a drawing to the sheet that changes the number of some of the sections, we only need to change numbers on the plan/elevation files and the Section Sheet file rather than opening each section drawing file to edit each title.

This practice is key to preventing CAD chaos. One principle involved here is to *keep the information dependency chain between drawings as short as possible*. That's why we use reference drawings: A change made on a building section will be instantly updated in the master "building section" sheet. A change made to the section designation can be quickly updated where it most matters—on the drawing that controls that information. Thus we somewhat reduce the danger of multiple versions of the building sections having different section number keys built into the independent drawing files.

Once again, we do not need to use Paper Space viewports in the final output drawing, since all sections are plotted at the same scale. We also gain an advantage over the *single drawing multiple view* model advocated for Paper Space because we have *multiple files* that can be worked on simultaneously by several people as the project schedule runs out of time—something impossible in the Paper Space model, where the massive file containing all the sections can (should) be edited by only one person at a time.

Using reference drawings introduces one danger that I would be negligent in not highlighting: There is no internal control in AutoCAD to prevent an outdated drawing reference from being retained in the file calling it. The new Reference Drawing manager dialog windows go a long way toward making this an easier task to monitor, but every project architect and job captain must be vigilant at all times as the final construction drawing set is assembled to check the content of the final sheet, especially if many people are working on the files that will make up the final output. This is really no different a management task than the one we have dealt with for centuries of manual drafting, except that the computer keeps more of the drawings hidden in invisible files rather than closed drawers or distant desktops.

Drawing Structural Framing Elevations Using the Building Sections and Elevations

The next set of drawings are relatively easily derived from the building section file we just completed, by erasing some objects and adding a few more. This type of drawing proceeds in roughly five stages, beginning with using the SAVE AS command to create a structural elevation file. Next we erase the interior walls, exterior walls, parapets, and notes which pertain to them. Also erased are the soil

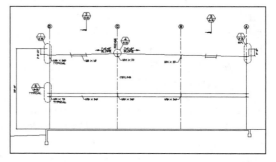

hatches and roof lines. The detail section keys were ⊕ MOVED up out of the way for possible future reuse. The first step in adapting the graphic elements of the section drawing was to cut the W8 roof member into a small indication of its size and orientation, TRIMming out the remainder.

Frame Member Type Indication

Next we gave the W18 frame support for the floor the same treatment. To show the frame member's orientation, we 🔲 OFF-SET the top and bottom lines 1 inch, both here and at the roof, to clearly indicate where the flanges on the beams are. The 🔲 LINE tool was used to draw Continuous lines over the column center lines; then Grips were used to shorten the column grid lines as shown in the illustration.

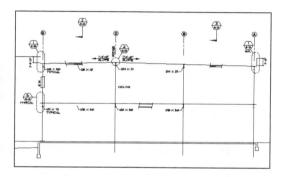

The 🔲 OFFSET tool was used again to indicate the orientation of the wide flange columns, and the Continuous column line was ⟋ TRIMmed out and redrawn as a Hidden line. In preparation for 🔲 COPYing this symbol to the other column locations, we used Grips to pull the lower ends of the other column lines down (270 degrees) past the floor slab.

We have now duplicated the wide flange column indication to all the column locations and have used the ⟋ EXTEND tool and the ⟋ TRIM tool to make the column line the correct height. A base plate indicator was added to the leftmost column line before we COPIED it to the other locations. As a last step, we ERASEd the previous three column lines by selecting their lower ends with a Crossing Box.

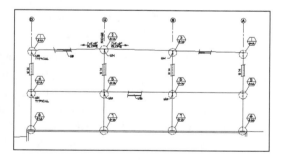

Final Annotation of the Structural Elevations

Frame joint detail annotation begins by using the 🔲 COPY tool to duplicate the ridge detail hex and circle as shown in the illustration. The 🔲 Edit Text (DDEDIT) tool is then used to type in the correct detail numbers and sheet reference (assuming we know what it is at this stage).

We also use the Edit Text tool to shorten the beam type designations to "W18" instead of the full "W18 x 50," for example. This isn't mandatory, just our office's standard. It's not terribly time consuming, and it sure beats erasing text manually. The other stylistic modification is to cut off the footings, but again, they could be left in the drawing.

Office Graphic Standards May Need Modification

Note that here's where office practices can be modified if you really want more drawing efficiency. Foundations aren't usually shown on manually drawn structural framing elevations, because they take time to trace. On a CAD system, they take time to erase! Keeping annotation consistent may be more important, because it reduces the possibility of confusion for someone unfamiliar with the drawings, and shorter text strings take less time to type.

Drawing Remaining Frame Reinforcing Members

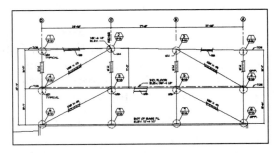

The final stage of the framing elevation creation is adding the objects not present in the original section drawing. As you can see in the illustration, we have added diagonal braces in the outer bays, first on the right side (because it is the lower elevation bay on the second floor), and we have used the MIRROR tool to duplicate them on the left side of the drawing.

Notes about the floor elevations are added with leader arrows. You can use the Leader Dimension tool and the Multiline Text Editor to do this fairly quickly and easily in Release 14. Experiment with it a bit, and you will find it may save you considerable time. The Help file will point you in the right direction on the exact steps required. Dimensions to the Top of Span and between column center lines are the last objects added, and the Elevation is complete.

Using the Correct Setting for the MIRRTEXT Variable

Another feature of AutoCAD is that it always defaults to making text read backward when text objects are MIRRORed. To avoid this problem, you need to set the MIRRTEXT variable to zero from its default of 1. Simply type <u>mirrtext</u> at the Command: prompt, then type <u>0</u> and press

[Enter]. Saving the drawing saves the variable setting as part of the drawing file. This variable setting should be saved in all your prototype drawings. Unfortunately, there is no way to make it a permanent setting within AutoCAD itself.

If you have text as part of a block, the text will be reversed if the block is MIRRORed, regardless of the MIRRTEXT setting. The only fix for this is to 🖉 EXPLODE the block before doing the MIRROR operation.

Drawing Floor and Roof Framing Plans

Again we use one of our earlier drawing files as the basis for a new drawing, this time renaming the Floor Plan of the second floor. The first stage of converting this drawing into a structural plan is to draw the beams around one of the stairs. We do this on a layer dedicated to structural steel. Next the ▢ BREAK In 2 points tool is used to cut each grid line at the building wall on both ends.

At this point, the ▧ Match Properties tool is used to change all the grid lines inside the walls to the structural steel layer and line type. Turning off all except the grid and steel layers helps considerably with this operation. The stair framing is next copied and/or rotated to the other stair locations.

The third stage of the process is using the ☁ OFFSET tool to create most of the secondary beams. Those that must be hand drawn are then added,

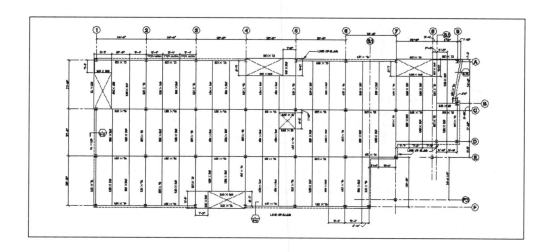

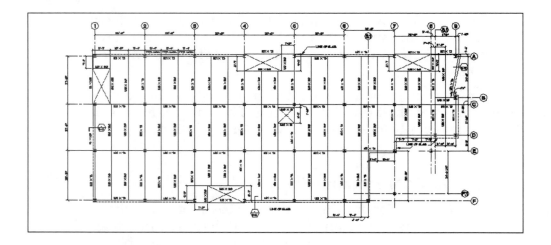

and one bay of the beam size designations is created using the DTEXT tool. These are then COPIED to every other bay using the Multiple option. The COPY tool is used repeatedly to place all the remaining steel type designations.

The final two steps require turning on the Bldgwalls layer and using the Properties tool (DDCHPROP) to place the inside building wall line on the Slab_edge layer that has a Hidden line type. Once the Bldgwalls layer is turned off, the FILLET tool is used to clean up and join all the corners.

Add dimensions and annotation, and the plan is complete. Note that where duplicate dimensions would be required for each bay of secondary frame members, we dimensioned one bay and called the dimensions out as Typical U.O.N. (Unless Otherwise Noted).

Strengths of CAD Drafting

The use of one drawing file to derive another drawing is one of the great advantages offered by AutoCAD and other CAD systems. In the previous examples, note how few of AutoCAD's hundreds of commands (tools) we actually used. We needed no third-party tools or software and created no Lisp programs or scripts. In the next chapter, we will explore how to take advantage of this strength of AutoCAD in a slightly different way when we explore Details, Schedules, and Notes.

CHAPTER SEVENTEEN

DETAILS, SCHEDULES, NOTES, LARGE-SCALE PLANS, AND ELEVATIONS

Contents

In this chapter we will discuss the appropriate use of AutoCAD for creating, storing, and retrieving details and other types of graphic information. Unfortunately many drafters and firms organize detail drawings in ways which make it difficult to keep track of detail drawing files, and burden the system with vast libraries of drawings which will never be used on future or current projects because the needed drawing cannot be located. The following examples will show you how to avoid these common mistakes and how to increase your detail production productivity.

Where the Devil Dwells

The devil is in the details. Mies was actually right about where God can be found, but souls are usually sold about a half to two thirds of the way into the production phase when we begin thinking "I'll do anything to get this project out to bid on schedule." God is only reliably found in nature's details.

The following illustration shows a detail sheet provided by a computer room floor manufacturer. It comes as a single large file (over 650 kilobytes) and is a tremendous help because it saves drawing effort and time. It shows a computer floor generic enough that it could be used to illustrate an "or equal" bid drawing. Nevertheless, it still needs to be broken up into a series of smaller files, each one containing a single detail.

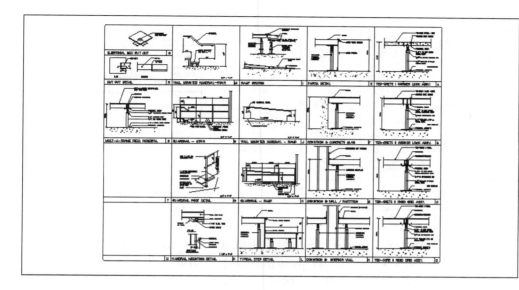

Here's why: If I am reusing the details on another project, I may not need all of them, such as the stair handrail. I may also need to move details around and insert new details in a different order. I and the whole office are better off with files we can pick and choose from. We can also make copies of individual details, modify them to suit the constraints of a particular project, and store the modification for later use. If individual details have their own files, it is easy to read a file name such as "Two_hour_partition_s-s_section.dwg" and know what the detail is about. A single file containing many details hides the information about those details, and you must go through the pain of opening each file and examining each detail pane to (maybe) find the drawing you want.

Don't Let Him Hide

Storing drawings as described in the previous sections forces the computer to mimic a manual practice that never worked very well in the first place. I have worked for companies that keep the details on paper only on project-specific sheets, tucked away in closed drawers or hanging on plan holders, their existence known only to the few and/or the departed. The best manual drafting firms keep bound libraries of details in 8½" x 11" format, accessible to everyone, and indexed in some method such as the CSI standard.

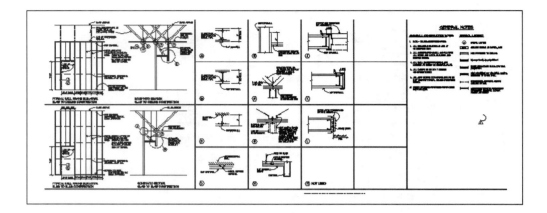

On the other hand, there are some drawings that almost never change and are as close to being "typical" as practically possible. The details illustrated here are our firm's common retail project interior framing drawings. Only the framing elevations and sections are drawn to the plotted scale; the rest are reference drawings stored in separate files and are inserted in the sheet, scaled to fit, dimensioned (dimensions are faked), and annotated on the sheet.

There are some invisible aspects to this drawing. The first is that there is no title block attached, ever. The second is that it is always plotted and pasted down on a title block sheet after irrelevant detail drawings have been Detached and new ones Attached. The General Notes section is modified as required for each project as well, and all text is DTEXT rather than MTEXT so that individual items can be moved and deleted in the drawing environment itself (we may change that as we gain more experience with Release 14's editor, or we will wait until AutoCAD delivers better note formatting capability).

Scale Issues: Professional Liability and Contract Control

There is a lot to be said for *not* plotting details to a defined scale (and making this explicit) on the final output drawing. This doesn't mean that the base drawings are not accurate, it just means that the plotted drawing is not to be scaled, and that contractors are therefore required to observe the dimensions shown. This gives the architect greater control over the division of responsibility for interpretation of drawings. Conflicts between small-scale drawings and large-scale details are more easily resolved when there is less leeway for "interpretation" on the job site.

The downside to this practice is that contractors often try to get dimensions from elevations or other drawings that are plotted to a defined scale, *and it places the burden of detail accuracy directly on the drafter*. If there are conflicts between the detail dimensions and the smaller scale drawings, blame has an easy target. No CAD system will ever change this relationship; it is noted here as a standing caution.

The CAD issue involved in plotting details to a defined scale (or not) is the amount of labor and time required to produce the physical document, including the amount of supervisory time spent checking details against the drawings they are keyed to. The impact of detail sheet assembly on CAD on the productivity of a firm is not so obvious and affects large projects more than small ones.

The Productivity Impact of Drawing Details on CAD Systems

There is a kind of Parkinson Effect that expands the time required to draw details on CAD systems by an exponent of the number of details required. The effect can be reduced with a library of predrawn details, but not eliminated. I call it a kind of Parkinson Effect because the original Parkinson's Law stated that work expands to fill the time allotted to it. In a perverse way, the time required to draw details on CAD systems seems to expand to a power of the number of details to be drawn, multiplied by the actual drawing time. For example, one detail may require one hour. Two details require two hours and thirty minutes, and the impact is obviously greater as the number of drawings goes up. The reasons for this are simple and somewhat obvious.

First, details are developed at the end of every project when time is short and hours worked eat into the project profit. Usually detail development exposes problems with specific parts of the building design; these problems must be fixed on smaller-scale drawings which were assumed to be "done."

Second, details are the last point at which the designer is required to make decisions about the items that the building's users may notice the most: lighting, hardware for doors and windows, paving and flooring, flashing and weather sealing (leak potential), etc. These are not decisions to be made in an offhand manner, and when every hour in a tight schedule counts, decision time impacts drawing time.

The use of CAD systems can only help these circumstances in that changes to the affected plans and other small-scale drawings are more easily accomplished on the computer than by hand. The actual creation of details requires a given number of people drawing drawings, period.

Sidestepping the CAD Detail Process Traps

If you are working on a normal drafting project as described in the above paragraphs, you need some way to deal with the pressure and push it ahead of you as much as possible during the design development phase. In our office we have found that detail development, even at the schematic level, can make use of the CAD system's geometry engine to speed up design decision making.

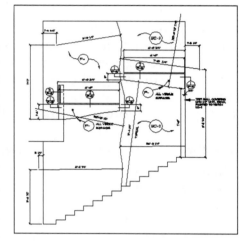

The elevation in the illustration was actually drawn as a design development study on AutoCAD, and dimensions and detail keys were added prior to plotting. Once the designer had the curves (arcs) placed as he wanted, we were able to simply dimension the radii and other features of the wall. The elevation was plotted and pasted on a detail sheet. Other details were hand drawn to complete the information required for the drywall and millwork contractors. Hatching was not used, partly because the project was a remodel of an existing building, and the contractor was responsible for verification of all existing conditions.

Here's a key to drafting fast track and/or complex design projects: The designers should be CAD proficient so they can work with the real geometry of the design on the CAD system. Failing that, drafters must start drawing key design geometry early, plotting it on small sheets and pasting it on paper title block sheets. Have electrostatic vellums made of the pasteups and submit prints for review. As I stated at the beginning of this book, creating detail development drawings as the project progresses is the best way to control the decision-making process and will help avoid destructive blame laying when you are out of time and information at the end of the project schedule.

Using the CAD System to Support Manual Detail Creation

The other way the CAD system can aid in detail development is for us to draw just enough on the computer to provide the foundation of the final detail drawing, which is completed by hand. The following example of a stair section shows how this works: We have taken the Stair_elevation schematic drawing that we used in the previous chapters and added more detail and dimensions that are quite efficient on CAD.

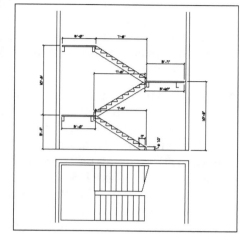

We will then plot the drawing, paste it to the detail sheet title block, and complete the annotation and draw the handrails by hand. Because hand drafting is competitive with **line by line** CAD drafting, we don't suffer a drawing time penalty, and we do gain the advantage of a complete paper document that can be printed and issued without further plotting.

• •

If You Must Use Paper Space: Drawing Setup

Paper Space, as we have noted in the early chapters, exacts some penalties on users and the organization of the firms using it. Judicious use of reference drawings can get around some of these limitations. In this example, we have opened a blank drawing file into which we have referenced (XREF Attach) fifteen detail files. AutoCAD will treat this as a single drawing in a Paper Space view, but other people can access and work on the details even when they are being referenced in an active assembly drawing.

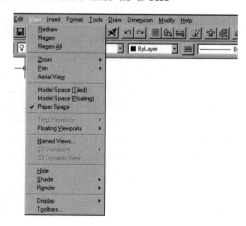

The next thing we need to do is to open our detail assembly drawing and Attach a title block as another xref, but not in model space. Instead, we will drop down the View Menu and pick Paper Space, then attach the title block drawing. By MOVING the title block out of the area of the original detail group, we can view the title block in Paper Space without making the other objects part of the final plotted drawing. Dropping down the View menu and selecting Paper Space and then regenerating the drawing gives us our work space and page size. The menu is what we will use to navigate between Paper Space and Floating Model Space viewports (this drawing file is included on the companion disk for you to experiment with).

Now we want to pick the Model Space [Floating] viewports option and drag a window for each viewport by using the 1 Viewport selection in the submenu. Repeating this operation gives us three views of the original drawing, as shown in the illustration. We have used the Paper Space selection from the View menu to draw lines around the detail area to guide the creation of viewports. Jumping back and forth between the Paper Space and Model Space [Floating] options may seem confusing, but it is fairly easy to navigate. Here's how to open a viewport and create a view of one of the detail drawings.

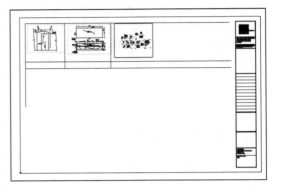

Note in the previous illustration that the active viewport is highlighted because it's the last one we created. As we create the viewports, they each show the same view of the entire base drawing. We have magnified parts of the detail drawing in the first two viewports and are now going to repeat the process. Here's a close-up of the viewport showing the complete group of referenced drawings. By picking the Model Space [Floating] option on the View menu and then picking the view-

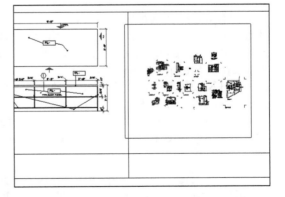

port border, we make the viewport active (the only current access to the underlying drawing). Now we need to pick the detail we want to focus on, and to select the scale factor we want it magnified by. This step is not intuitive but is easy nonetheless.

This viewport is going to show just one of the details in the base drawing, but for now we need to stay in Paper Space and ZOOM in to the close-up of the viewport shown in the illustration. Then we can switch to the Model Space [Floating] view and pick the viewport border to make it the active one. To magnify the detail to scale we use the 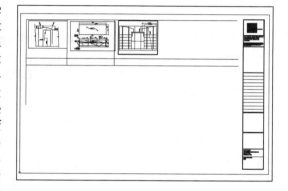 ZOOM Center tool. This may involve some experimentation and use of the UNDO [U] tool. Select the ZOOM Center tool from the Zoom tool bar and then place the cursor on the detail to be magnified. Pick the visual center of the elevation in the upper right corner of the group (it looks like the largest rectangular blob in the illustration).

Selecting Scale Factors for Details in Paper Space

Starting from the scale factor for the base drawing (full size with dimension text scaled to plot at defined magnification factors, which we will address in the next section), we need to apply a magnification number for the details. The easiest way to think about this is to use the smallest plotted scale as the baseline. If we are plotting all our plans at 1/8" = 1'-0", then we can use this as a base for scaling drawings up (relatively). If we plot our base Paper Space drawing at 1/8" = 1'-0" then we need to scale our details by factors of 8 (10 if you are fortunate enough to draw in the metric system). In this specific example, we will scale the elevation by a factor of 12, which will plot at 3/4" = 1'-0".

Here's how to do the magnification: Pick the ZOOM Center tool from the Zoom tool bar, and pick the approximate visual center of the elevation. Now type 12X on the command line, press [Enter], and the elevation will be scaled to 3/4" = 1'-0". Once you have the view in the window, you can use the Dynamic Pan tool to move the detail into a precise position. The magnification factor is easy to figure out. Pick up any drafting scale and count the number of 1/8" divisions on any of its scale sections. The count gives you the scale factor. The *x* character tells AutoCAD to scale the drawing by the number entered by the original full size of the base. In our example, you can now see how the elevation has been scaled in the viewport.

The final steps needed to make the detail sheet plot ready is to add border lines (as we are starting to do) by dropping down the View menu and picking Paper Space. Now you can draw the lines and add scale notes and detail titles. The last step is to make the viewport borders invisible. I like to put them on a dedicated layer and turn that layer off (possibly because I'm a control freak), but you can also put them on the Defpoints layer, that though visible, never plots. As you can see in the illustration, the detail sheet now looks as it should, ready for plotting.

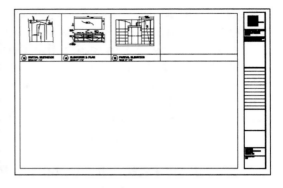

Why Paper Space Makes Dimensioning Easier

Dimensions can be omitted from the base drawings and added in Paper Space with the Model Space [Floating] option on the View menu selected. You must open your Dimension Style and set the text and tick mark size appropriately to the scale magnification used in your ZOOM factor, but the dimensions will be accurate. If you keep the dimensions in Model Space, you can create various styles based on the anticipated ZOOM scale factors so that you get consistent text and tick sizes in the Paper Space views.

All of this is much more desirable than struggling with Model Space drawing scale factors that change the value of your Associative Dimensions or make giant text and tick marks in Nonassociative Dimensions as drawings are scaled up.

Paper Space Negatives

The previous examples are a seductive. Paper Space views are exceedingly easy to create in Release 14. There is a downside to all this: You must still plot a sheet of details at some point. You can plot individual drawings if you develop them separately as I suggest, but because details use lots of text and hatches (even with R 12's better speed in handling these objects), they take virtually forever to plot. Because some parts of the final drawing exist only in Paper Space, check plots of the full sheet are hard to avoid, even when making small changes.

Incremental development of detail drawings as individual files that are plotted when complete and then pasted up provide much better drawing development control and error checking. This is especially critical when cross-checking detail references in other drawings. It is virtually impossible to check detail references on several drawings against a detail assembly drawing solely in the computer.

Finally, remember that being able to mix detail drawings created on paper, started in CAD and completed on paper, and CAD-drawn details pasted up on title block sheets gives a production team ultimate flexibility in dealing with the inevitable time crunches at the end of a project.

Schedules and Notes

All of the above comments on details apply even more for schedules and notes, which can fill a complete sheet. When I first got hold of Release 14, I jumped into producing a 36" x 48" sheet of Multiline Text notes using True Type fonts. The monster took almost 2 megabytes of disk space and over thirty minutes to plot. I tried to bring it up in Release 13 and 12, and it crashed both times.

After numerous experiments, I must conclude that we are still much better off typing door schedules and other complex or voluminous notes in a word processor or spreadsheet and pasting up printouts on quickly plotted title blocks. I have timed the whole process, and though aligning text sheets in pasteup on paper is not my favorite thing to do, by the time I spent several hours more completing one sheet of notes using MTEXT, I decided I'd stick with pasteups for now. This situation has not gone unnoticed by Autodesk, and the next Release of AutoCAD may offer relief for this problem.

Large-Scale Plans

Plans of complex areas such as rest rooms and public spaces in hospitality projects must sometimes be drawn on sheets dedicated to larger-scale representation because of the amount of annotation required for elevation keys, keynote references, material and finish keys, complex floor paving patterns, etc. The usual scale for such plans is ¼" = 1'-0".

There are several approaches one can use to create these drawings on a CAD system, and each has its advantages and drawbacks.

Individual Drawing Files That Are Referenced by Small- and Large-Scale Plans

This is strictly a CAD-centric practice. You actually draw the large-scale plan full size, just like other drawings, but in its own file. It has no dimensions or key references at all; it is just the basic geometry that other drawings will annotate. It should also have no text that would create scalability problems in other drawings. This file is then Attached as an xref by the files it will be actually plotted on, such as the Floor Plan drawing and the large-scale plan drawing.

The advantage to this approach is that any changes made to this drawing will show up in all the "sheets" calling it as a reference drawing. The plan file can also call the Floor Plan file as an Overlay xref so that you are not working in isolation from other parts of the structure when making changes. Because the file is small, it can contain layers for FF&E (Furniture, Fixtures, and Equipment) objects, flooring, and millwork, which can be turned off and on by the drawing calling the reference.

Elevations are easily developed off of this type of plan drawing, just as we did for the building elevations in the previous chapters. The drawing calling the large-scale plan can also call the elevations' file, using different insertion points, and xref Clip helps in masking out unwanted parts of the same file.

One disadvantage to this method is that you need to open another file just to make minor changes that may affect only one file such as the Electrical Power Plan in the Interior drawings. Another is that you must create a distinct Dimension Style with a scale factor set to the magnification used by the plot scale of the drawing calling the dedicated plan file as a reference. If other dimension scale factors are used in the same file (for instance, on elevations and larger-scale details to be plotted on the same sheet), then we have introduced a potential for serious and hard-to-detect errors to occur. The wrong dimension child style can easily be used on a drawing by someone not paying close attention to what he or she is doing (I have even seen this happen in manual drafting!).

Using Paper Space to Create Different Scale Views of the Basic Floor Plan File

Using the general Pspace construct we described earlier for detail assembly, you would ZOOM Center into the Floor Plan file and use the Paper

Space viewport to clip out the part of the plan to be scaled up. For a ¼" = 1'-0" plan, the ZOOM scale factor is 2. By going into Model Space (Floating), you can dimension and annotate to your heart's content by using dimension styles set up for your Pspace magnification.

The downside to this use of Paper Space is that the Model Space Plan and Paper Space drawings must be the same file, which limits how the file may be used by other drawings calling the Model Space part of the file as a reference. This introduces an extra layer of complexity to the CAD project structure that may be more work than it's worth. There is also the need introduced to know (and remember) which information is contained in the Model Space view, and which is only in the Pspace view(s), none of which is obvious when looking at one or the other on the screen.

Combining Paper Space and Xref Drawings as a Good Compromise

Taking the approach we did in the detail sheet described earlier in this chapter allows us to use a relatively small Paper Space drawing view to "call" several files. Using different insert points, we can leverage the first approach we described in the independent file concept for drawings that need to be plotted at different scales. Suppose that we want to combine our rest room plan file and rest room elevation file with details drawn in Paper Space. We simply call these files as xrefs to our separate Pspace drawing file (with separate insert points so the drawings don't overlap in Model Space) and create viewports for each part we need to see. Our drawing file can then contain drawings of the *details* and keys to them entirely within its own environment, scaled as we want. Or we can extend this model to call a stand-alone detail drawing file as an xref (just like in the earlier example on detail sheet assembly), from which we can select views needed.

The model can be extended even further to use independent detail files called as xrefs with different Model Space insertion points. We thus have almost no overhead invested in our "sheet" because it just references all the drawings we have created during the production process, assembles views of them, and adds a title block and drawing designation. The following illustration shows how this can be used to create different views of two drawings: the base building plan and an elevation of the restroom from a completely different drawing file.

This is where Paper Space offers a distinct alternative to pasting up separately plotted paper output. By referencing many files into one final sheet with different scale views of each file, we have overcome some major problems in assembling different scaled drawings, especially late in the production process. Many different people can work on the referenced files at the same time, and at plot time the "Stop drawing!" order can take an inclusive snapshot of the drawing process.

Physical Pasteup of Independently Plotted Files

Now we will consider our last alternative to the models just discussed. This is the "low-tech" method of CAD drawing management and production, and as we have said before, it has its advantages. There are really two parts to this method of working, which I have talked about in previous chapters. First, we can use a cascading pattern of reference drawings, shifting between plan, elevation, and section during both design development and construction drawing assembly to keep dependent and independent drawings referenced to the same core geometry.

Once we have reached the detail drawing stage, we can use independent files such as a rest room plan and an elevation drawing file, or a rest room plan and an independent rest room elevation file(s), as complete drawings that are to be plotted without title block and pasted up on paper for eventual final output. It is even useful to plot each elevation of such a file separately as it is developed, so it can be checked during the drawing process.

Now we have a paper version of the CAD drawing in process, which can have changes made to it piecemeal (which is how most corrections to drawings are made) without replotting the entire sheet. At the end of the project schedule, there is no need to replot every detail and multi-scale sheet! Just print a final electrostatic vellum copy of the pasteups and start printing bid sets.

This method can save the small to medium architectural office many hours of plot time. It will have a small impact on the largest firms because they have already set up strictly defined production output processes and are networked to high-speed plotters. They can presumably afford to waste (recycle) paper. However, even large firms can benefit from the reduced check time cycle afforded by this method of assembling and reviewing/revising drawings at the end of the production schedule.

Using the Most Efficient Methods, Even if They Defy the Oxymoron of "Conventional Wisdom"

Those are the broad-brush alternatives. To some degree, the use of Paper Space has become a religious issue (you have to believe it saves time and reduces complexity as opposed to knowing via time tracking that it really does), and the nonbelievers are castigated as heretics and technophobes. It's your choice. If I think my firm can outproduce your firm and underbid your production fees on any project we are competing for, who gains an advantage? If you have to lower your profit to get the project, how long do you expect to survive if a recession hits?

The Devil Is Alive and Well

The Business of Drawing today is somewhat distinct from the Practice of Architecture, and that's too bad. We are in a position to restore the union between design drawing and the final representation of it by drafting, but only if designers can bring themselves to deal with the challenges posed by CAD drafting and to use this powerful tool to aid in translating vague visions into dimensions and details. Never in the history of building have we been able to draw structures of any size with such accuracy. The potential of CAD systems today to contribute to the design and profit of architectural firms is limited only by the people who manage and use them.

CUSTOMIZING TOOL BARS AND TOOL BOXES

Contents

Adding a Tool Bar to the Drawing Area

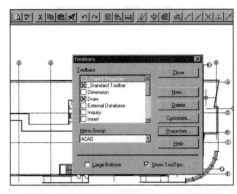

We're going to add the Object Properties tool bar to the Standard tool bar. To add any tool bar or box to the drawing area, right click on a tool in any tool bar to bring up the Toolbars dialog window, or pick Toolbars from the **View Menu** if none are visible on the screen, to open the dialog.

With the dialog box open, pick the Object Properties check box to highlight the name. The tool bar will immediately display at the top of the screen, which is its default position (we'll cover moving and resizing tool bars and boxes shortly).

Adding the Modify and Draw Tool Boxes

Scroll down the list of tool boxes (bars) and find the Draw and Modify names. Check their boxes as well and then pick the Close button. You can now drag the Draw and Modify tool boxes to the positions you want on the screen. If you drag a tool box to the top, bottom, right, or left edge of the screen, it will automatically become "docked" (be combined into the drawing screen border). Otherwise it will "float" on top of the drawing.

Customizing Tool Bars and Tool Boxes

We are going to add a new tool, the Xref command button, to the Standard Toolbar at the top of the screen. The steps outlined below are the same for adding any tool to any tool bar or box (or removing any tool).

Adding the 🖫 Xref Tool to the Standard Tool Bar

To customize any tool bar or box, start by doing the following: Right click on a tool in the tool bar to be customized to bring up the Toolbars dialog box. Pick the Customize button on the dialog. A new dialog titled Customize Toolbars will appear with a drop-down list and a few obscure icons below it. Pick the drop-down list and scroll down to the type of tools you want to use, and pick the name on the list. In this case we will pick **External Reference**.

A group of tool icons will appear for you to select from. To display the name of any individual tool, just pick the icon; a cryptic description of what the tool does and the Auto-CAD command it invokes will be displayed in the Description section of the dialog box.

As you *cannot* see in this illustration, we have picked the tool icon in the lower left corner of the group and have discovered that it is the one we want. The description reads "External Reference: Controls external references to drawing files: xref." The command, XREF, is listed last, after the description. While the description is technically correct, it would have been more helpful if it told us that the tool "opens the External Reference Control panel for accessing reference drawing files." Oh, well.

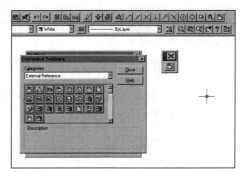

To add this tool to the Standard Toolbar, simply pick it and hold the mouse button down while you drag it to the position desired on the tool bar, and lift your finger off the mouse button to drop the tool in place. In this illustration, we have dropped it at the right end of the tool row, next to the Running Osnap tool. Your Standard Toolbar will probably not look like the one in the illustration, because we have already dragged all the Osnap tools from their group in the Customize box up to the position you see illustrated.

Adding Floating Tools to the Drawing Screen

If you drag the XREF tool icon to the drawing screen and drop it there (as we have also done), a mini tool box will be created with just the one tool in it. You can leave just the single tool or add others by dragging and dropping their icons onto this new tool box. Picking the X at the top of the tool box will make the box disappear. This is a good way to quickly drop a tool or a small group of tools that you need temporarily for short, repeated use, but that you don't need cluttering up your screen space for long periods of time. Unfortunately, you cannot save these temporary tool boxes as part of the AutoCAD tool list. To save them with your own tool box design, read on.

Creating Custom Tool Boxes

A variation on this procedure easily allows you to create new tool bars and boxes. Go back to the Toolbars dialog by pressing the **Close** button on the Customize dialog box. Pick the **New** button on the Toolbars dialog. A new dialog box will appear with a window to type your tool box name in. Type it and pick the OK button on the dialog. Unless you are an expert at AutoCAD menu customization, leave the menu drop-down list at the bottom of the New Toolbar dialog box alone.

The New Toolbar dialog box will disappear and a blank tool box will be displayed floating on the drawing screen. Pick the Customize button on the Toolbars dialog window again to open the Customize dialog. Just use the drag-and-drop technique we described earlier to populate your new tool box with icons of your choice. Now if you close your new tool box (by picking the X button), it will appear on the Toolbar dialog drop-down list, which you can access at any time just by right clicking on any tool on the screen.

Customizing Tool Bars and Boxes by Moving Tools from One Location to Another

I like to have the Tracking tool located next to the Line tool in the Draw tool box. Unfortunately, Tracking is normally part of the Osnap tool box, which I use less frequently (I rely on Running Osnaps because Autosnap makes them so easy to manipulate). As long as you have the Customize dialog on the screen, you can drag any tool on any tool bar/box to any other tool location.

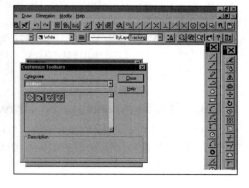

Getting Rid of Tools You No Longer Want in a Tool Bar/Box

As the illustration shows, we now have the Tracking tool both on the Standard Toolbar and on the Draw tool box, a situation that would occur if we dragged it from the Customize dialog to the Draw tool box. To remove the duplicate tool on the Standard Toolbar, simply drag and drop it onto the Customize dialog box. This method works for all tools on all tool bars and boxes.

Changing a Floating Tool Box's Shape and Location

Depending on how you have set up your drawing screen, you may need to move tool boxes around to see a part of the drawing they cover or to reshape them for the same reason. To move a tool box, simply pick the border of the box and hold the mouse button down while you drag the tool box to another location.

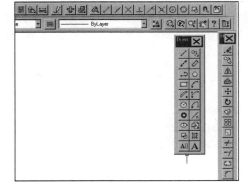

To change the shape of a tool box, pick the border (called the "grab region" in ACAD-speak) and hold the mouse button down until a small double-headed arrow appears. As you drag the mouse toward the tool box or away from it, you stretch or shrink it in the direction you move the arrow. Getting the desired shape may take more than one try, but don't be afraid to experiment.

Saving Your AutoCAD Customizing in a Personal Profile

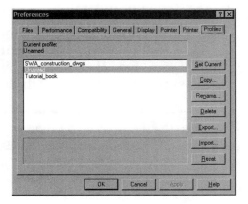

If you work in a company where someone else will be using your workstation while you are on vacation or traveling, you can preserve your work environment customization by saving it as a personal profile. From the **Tools menu**, select **Preferences** to open the Preferences dialog window. Pick the **Profiles tab** at the top of the dialog, as shown here.

Creating a New Profile Name

Now for the counterintuitive part: To create a new profile name, pick the **Copy** button. That's right. Don't ask why, because I haven't a clue. Just pick the button. Type your profile name in the window, and then pick the **Set Current** button to make that your current profile for the drawing session. That's all there is to it. When you exit AutoCAD, all your settings will be automatically saved to this profile name.

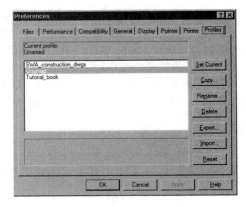

Advantages to Using a Personal Profile

When you leave your workstation for long periods of time, go back and set the current profile to Unnamed, and your own setup will be protected from anyone who comes in and messes with your tools, your screen background color, etc. If someone else regularly uses your computer, he too can set up a personal profile on it so he doesn't interfere with your settings.

You will notice the **Export** and **Import** buttons. Because AutoCAD saves profiles as part of the Windows 95/NT Registry instead of the old acad.ini file, you can't copy a specific file to move your personal profile to another machine. This is actually a good thing because copying the acad.ini file to someone else's machine in the past would overwrite all of that person's configuration settings! Make sure you Export to a floppy disk to make the process easier, and read the Help file before doing so. If you think this is a gray enough area to cause you to be apprehensive, get your system administrator to help you the first time through.

Import and Export are particularly valuable if you are moving permanently to another computer or upgrading your system to a new computer. It removes the tedium of going through all the Configuration processes required by older Windows versions of AutoCAD.

Warning: There is no way to Export your profile to AutoCAD Release 13. If you try to do so and R13 is running on Windows 95 or NT, you could mess up the Windows system Registry on the target machine. This is not a bug, but the application of DDT (Don't Do That!) can prevent problems.

Don't Be Afraid to Customize the Way AutoCAD's Tools Work for You

With the introduction of the Windows environment, I have seen numerous AutoCAD users resembling deer caught in oncoming headlights, mesmerized in front of their monitors, afraid to change anything on the new and unfamiliar screens. Don't be afraid of crashing Release 14 by modifying its tools and tool bars. There are ways to crash the system, or at least make it behave badly, but customizing tool bars is not one of them.

In fact, Release 14's tools are so easy to change that I encourage you to try many different combinations and to try tailoring your tool set to particular projects. You can never be sure what will work best for you until you try it, so spend some time experimenting to find the best ergonomic configuration for the way you work.

DRAWING CEILING AND TILE GRIDS THE QUICK AND EASY WAY

Contents

We are going to explore a general method of generating grid patterns for floor and ceiling grids using the ▦ Hatch (BHATCH) tool. Introduced in Release 13 and vastly improved in 14, this tool allows us to quickly create precise grids within complex areas, align them, TRIM and EXTEND their edges with Fence selections, and go on with the rest of our work. Before automatic hatch boundary selection came to AutoCAD, we had to trace a Polyline around an area or laboriously pick individual objects to form a boundary.

It was a toss-up as to which method was slower (neither was fast)—HATCHing or ARRAYing lines. Now all we need to do is pick one point in the area we want to hatch and AutoCAD does the boundary-building work for us.

In this short tutorial, we will create a ceiling grid on the first-floor A-1 _base.dwg file. Once we have completed all the ceiling grids, we will turn off all other layers, make the ceilings into a Block, and WBLOCK them to their own Ceiling Plan file. We would then open that file without saving the changes to A-1_base.dwg, and Xref A-1_base.dwg into our ceiling plan file. Floor Covering Plans are created the same way. The basic technique follows.

● ●

Creating the Horizontal Grid Lines

We have created a Ceiling_grid layer and made it Current, and we are ZOOMed in on the lower right space just off the entrance atrium. The Bldgrid layer is turned off because it would break up the hatch.

We have picked the ▦ Hatch (BHATCH) tool from the Draw tool box. The illustration shows the Boundary Hatch dialog box floating over our view of the space. We have selected the Line hatch type. This is the most common AutoCAD standard hatch pattern used in creating diagonal and orthogonal grids, because only it allows different angle and distance offsets between the "horizontal" and "vertical" lines (more about that at the end of the tutorial).

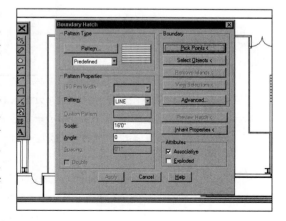

Setting the Spacing for the First Grid Lines

In the dialog box we have left the angle setting for the hatch at 0 degrees, which is horizontal to our drawing. Notice, however, that we have set the hatch spacing (labeled **Scale:** in the dialog) to 16'-0". This hatch draws lines spaced ⅛" apart when the spacing is set to the default of one inch. That means that 8'-0" = 1'-0" in scaling this pattern. 2'-0" therefore requires a 16'-0" spacing. Lines 1'-6" apart would equal 12'-0" in the hatch scale.

Creating the Hatch Boundary

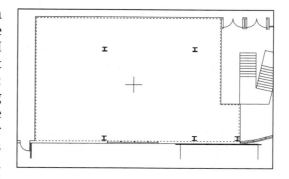

The next step is to select the hatch boundary. We do this by picking the **Select Points** button on the dialog. I think this button should be named Select Point, but what do I know? We'll see why: When the button is picked, the dialog box disappears, and we are back at the unobstructed plan view without a clue or a prompt telling us what to do. All that is required is to put the pointer roughly in the center of the space to be hatched and press the mouse's pick button.

AutoCAD then finds the logical (!) boundary formed by the objects nearest to the point picked. It will automatically (unless instructed otherwise) isolate text and closed objects within the area and will intelligently hatch around them, leaving a margin at text objects, but hatching in contact with most others. Once the boundary is complete, it is shown highlighted on the screen as shown in the illustration above.

You now have no option but to **return** to the Boundary Hatch dialog box, but again you are presented with no clue about how to do that. It's really simple: Press the Mouse's Enter button, or press the [Enter] key (once known as the Return key). With the dialog box back in view, we can select **Preview Hatch** or simply pick **Apply** if we are satisfied with the selected boundary. If there was a problem with the boundary created, you may need to pick the **Select Objects** button and individually pick lines and objects (you can use any selection method including Wpoly and Fence) until AutoCAD gets it right and/or stops giving you error messages. You know you have an area with leaks if there is a long delay and you finally get a message like: "12,345 objects found. Do you really want to do this?"

Once the Apply button is pressed, the hatch is created, as shown in this illustration. We now have a horizontal grid of lines 2'-0" on center filling the space. Our next step is to create the 4'-0"-on-center grid running vertically.

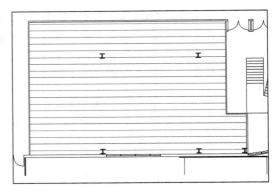

Pick the 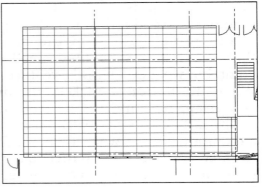 Hatch (BHATCH) tool one more time, or press the mouse's Enter button to recall the BHATCH command. In the dialog we make two simple changes: First we set the hatch spacing (Scale) to 36'-0", which equals 4'-0" on center. Then we set the hatch angle to 90 (degrees) and pick **Apply**. Just that fast we have a 2' x 4' ceiling grid.

Aligning the Grid within the Space

There's only one problem: The grid is not where we want it (in this case, centered on the column grid lines). AutoCAD once again mysteriously failed to read the designer's mind! To quickly fix this situation, we turn off all wall and column layers and turn *on* the Bldgrid layer.

Next we pick the 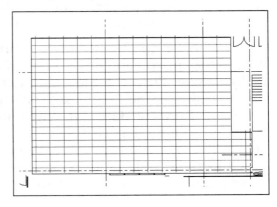 MOVE tool on the Modify tool box, and use a Crossing Box to select the vertical and horizontal hatches (you need to cross just two lines to get everything). We select one of the **Intersections** of the hatch grid and move it **Perpendicular** to a vertical or horizontal building grid line. We repeat the process for the next direction.

Now we must use the EXPLODE tool on the Modify tool box to turn the hatches into individual lines. With the grid aligned we will proceed with the EXTEND and TRIM operations by turning off the Bldgrid layer so we don't trim that also.

Picking the ◢ TRIM tool from the Modify tool box, we pick the hatch boundary as the trimming object, then use a Fence to cut all the ends of the lines projecting beyond the top and left side of the hatch boundary. You may need to ZOOM in to start each Fence line accurately.

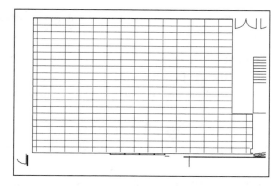

Next we will use the EXTEND tool on the Modify tool box to complete the grid. First pick the hatch boundary as the line to Extend to. Then draw a Fence line vertically through the right-side horizontal lines, followed by a Fence line drawn horizontally near the lowest side of the grid, through the vertical lines.

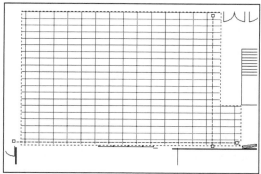

That's all there is to it. We're done with this grid. All we need to do is turn on the wall, column, and Bldgrid layers and we have the basis for a Reflected Ceiling Plan. Notice that I said the *basis* for a reflected ceiling plan. A bit of work needs to be done before we can begin to add lights, HVAC supply and return grills, ceiling type designations, and all the other objects such a plan will eventually possess.

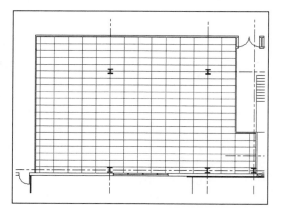

Getting from Grid Pattern to Reflected Ceiling Plan

Once all the ceiling grids are completed, all layers except the grid layer itself are turned off or frozen. I even put the hatch boundary Polylines on a separate layer and let them stay as part of the A-1_base drawing so I can use them for area calculations later on. Then I create a Block out of the grid and WBLOCK it to the A-1.X_ceiling.dwg file.

This new drawing file, with only the ceiling grid lines on it, is opened and the A-1_base plan is attached as a reference drawing. Now the process of adding door headers, lights, and other objects can begin.

Drawing Grids on Angles Other Than 0 Degree and 90 Degrees

To rotate a grid at a specific angle, simply enter the angle for each hatch component in the Boundary Hatch dialog box. If we had wanted to set our ceiling up as a 2' x 2' grid rotated 45 degrees to the vertical, we would have entered 45 for the first angle and 315 degrees for the second hatch application.

Virtually any angle can be used, but you will need to LIST objects that are at unknown angles with which your grid may need to align, and then type the exact value in the dialog box. Remember to always explode the hatches, even if you don't need to do it to TRIM or EXTEND them after repositioning a pattern. Hatches, even with Release 14's better management of them, still take up a huge amount of memory and disk space. Especially on large drawings, I explode every hatch I can get away with, and I always keep them on dedicated layers to make editing them easier.

Creating Floor Tile or Stone Grids

Floor grids are created exactly like our ceiling examples above. Even the floor covering or hardscape paving plan can be created by the same methods, using exactly the same steps as for ceilings. The only difference is that you must add space in the grid for the size of the grout joint to be used. For example, a stone floor of 12" x 12" tiles with a ⅛" grout joint would be entered in the Scale box as 8'1" because the scale base of the Line Hatch is 1" = ⅛" of separation between lines.

Any tile surface can be hatched this way, including tile on walls in elevation drawings, as long as the tile is laid in a regular grid. Grids used to be a pain to do, especially before Release 13. Now they need not be feared as a design element or a drafting problem as much as in the past.

ABOUT THE DISK

Contents

• • • • • • • • • •

Introduction

The enclosed disk contains files of drawings discussed in this book. The files were created and saved in AutoCAD Release 14. In order to use the files on the disk you need to have AutoCAD Release 14 or AutoCAD LT 97 installed on your computer. Please refer to the readme text file for the disk table of contents.

The drawings are yours to modify and use for your own purpose in your work, with the exception of duplicating the Tutorial Building as an actual built work. You are free to extract any component or part of the Tutorial Building you need and use it on a real project. Tip: Adding jamb lines to the door blocks and setting the insert point to Endpoint of one of the wall lines may save you some time. Experiment!

Minimum System Requirements

- IBM PC capable of running AutoCAD Release 14 or AutoCAD LT 97
- Windows 95, Windows NT 3.51, or Windows NT 4.0
- AutoCAD Release 14 or AutoCAD LT 97
- 3.5" floppy disk drive

How to Copy the Files onto Your Computer

The files on the disk can be copied to your hard drive using the standard methods supported by Windows Explorer. To copy files to your hard drive, go to Windows Explorer and create a new directory on your hard drive. Select all the files from the floppy disk drive A and copy the files to the new directory you created on your hard drive.

User Assistance

If you need basic assistance or have a damaged disk, please contact Wiley Technical Support at:

Phone: (212) 850-6753

Fax: (212) 850-6800 (Attention: Wiley Technical Support)

Email: techhelp@wiley.com

To place additional orders or to request information about Wiley products, please call (800) 225-5945.

INDEX

For information about the disk, refer to pages [323–324].

AutoCAD and AutoCAD LT are registered trademarks of Autodesk, Inc.
IBM is a registered trademark of International Business Machines Corporation.
Microsoft, Windows, and Windows NT are registered trademarks of Microsoft Corporation.

Learning Resources
Centre